Techniques of Differential Topology in Relativity

ROGER PENROSE
Birbeck College
University of London

Techniques of Differential Topology in Relativity

SOCIETY FOR INDUSTRIAL AND APPLIED MATHEMATICS
PHILADELPHIA

10 9 8

ISBN 978-0-898710-05-2

Acknowledgments

The material published here was based, in part, on notes taken by D. Lerner at Regional Conference lectures, July 13–17, 1970, delivered by the author at the University of Pittsburgh. The work was supported by the National Science Foundation under Grant GP-18904.

ROGER PENROSE

Preface

The purpose of these notes is twofold. In the first instance, it is to acquaint the specialist in relativity theory with some modern global techniques for the treatment of space-times. It is hoped that the detail given here will be sufficient to enable him to use these techniques when needed and perhaps to incorporate them into his way of thinking. Secondly, it is intended that the notes may provide the pure mathematician, who has some knowledge of differential geometry, with a way into the subject of general relativity, so that he may be able directly, and without detailed physical knowledge, to employ his mathematical understanding and special insights in a field which does have a deep interest for physics.

The scope of the notes will be the mathematical background necessary for a detailed comprehension of the proofs of the so-called "singularity theorems" associated primarily with the name of S. W. Hawking. Also some of the related body of knowledge which has grown up in association with these results will be covered (see [1]–[11], [18], [21]–[32], [34]).

The standard of rigor adopted will, I hope, be adequate. Where arguments are not spelled out in complete detail, it should be fairly obvious how these details may be supplied. But on the whole, I have gone into rather more detail here than is to be found in other works on this topic. Some of the basic results have had something of the status of "folklore theorems," the proofs of which had not, to my knowledge, been spelled out before. It is my hope that these notes may be able to remedy this situation to some considerable extent.[1]

A basic knowledge of point set topology will be assumed; also the essentials of (intrinsic) differential geometry, according to *either* a "modern" *or* a "classical" point of view, will be needed. The notation used, if not always totally conventional, will, it is hoped, be tolerably acceptable to both classes of reader. In the occasional places where a tensor formula needs to be employed, I shall normally give a parallel treatment using both the "modern index-free" and "classical kernel-index" notations. This will enable familiarity to be gained by those at ease with only one of these notations—and will emphasize that the difference is essentially only a notational one. In fact, if desired, the "kernel-index" expressions can always be read in accordance with certain conventions whereby indices are to be interpreted as abstract labels and indexed symbols are interpreted in an abstract coordinate-free manner [9].

[1] There is, in addition, a forthcoming book by Hawking and Ellis to be published by the Cambridge University Press. This will also cover more of the same sort of material from a slightly different viewpoint.

Although we shall be concerned primarily with the case of a four-dimensional space-time (hyperbolic normal signature $+, -, -, -$) the entire discussion applies equally well to "space-times" with any positive number of space-dimensions[2] and one time-dimension (that is to say, to any time-oriented hyperbolic normal pseudo-Riemannian manifold of dimension two or more). Indeed, many of the illustrative examples will be two- (or three-) dimensional. It is possible, therefore, that the ideas discussed here may find application in contexts other than those primarily intended, for example, to the general theory of partial differential equations.

ROGER PENROSE

[2] The discussion of conjugate points on null geodesics becomes vacuous, however, unless there are at least two space-dimensions (cf. Section 7).

SECTION 1

Preliminaries

1.1. DEFINITION. A *space-time* M is to be a real, four-dimensional[1] connected C^∞ Hausdorff manifold with a globally defined C^∞ (or C^2 would do) tensor field g of type (0, 2), which is nondegenerate and Lorentzian.[2] By *Lorentzian* (or hyperbolic normal) is meant that for any $x \in M$ there is a basis in $\underset{x}{T} = \underset{x}{T}(M)$ (the tangent space to M at x) relative to which $\underset{x}{g}$ has the matrix diag $(1, -1, -1, -1)$.

1.2. DEFINITION. Let M be a space-time, with[3] $x \in M$. Then any tangent vector $X \in \underset{x}{T}$ is said to be: *timelike*, *spacelike*, or *null* according as $\underset{x}{g}(X, X) (= \underset{x}{g}_{ab} X^a X^b)$ is positive, negative or zero. The *null cone* at x is the set of null vectors in $\underset{x}{T}$. The null cone disconnects the timelike vectors into two separate components.

1.3. DEFINITION. A space-time M is said to be *time-orientable* if it is possible to make a consistent continuous choice all over M, of one component of the set of timelike vectors at each point of M. To label the timelike vectors so chosen *future-pointing* and the remaining ones *past-pointing* is to make the space-time M *time-oriented*. In this case, the nonzero null vectors are termed *future-pointing* or *past-pointing* according as they are limits of future-pointing or past-pointing timelike vectors.

1.4. *Remark*. A space-time is clearly time-orientable if there exists a nowhere vanishing timelike vector field. The converse is also true. This follows from general theorems on the existence of cross-sections of fibre bundles (the fibre consisting of future-pointing timelike vectors at a point being "solid"; cf. Steenrod [12]). Alternatively, we can construct a nowhere vanishing timelike vector field on a time-oriented space-time M by using the fact that M can be given a positive definite Riemannian metric h. Choose the vector field V as the unique future-pointing unit eigenvector with positive eigenvalue λ, of g with respect to h. (That is, $(g_{ab} - \lambda h_{ab})V^b = 0$, $h_{ab}V^aV^b = 1$; i.e., $g(V, W) = \lambda h(V, W)$, $h(V, V) = 1$ for all vector fields W.)[4]

1.5. *Side remark*. For a given abstract manifold, the condition that it admit a time-orientable Lorentz metric is the same as the condition that it just admit a

---

[1] As stated in the introduction, although explicitly the arguments refer only to four-dimensional space-times the results will all extend in an obvious way to a space-time of n-dimensions, $n \geqq 2$.

[2] Such a manifold is necessarily paracompact (see Geroch [10]).

[3] I adopt the usual slight abuse of notation here.

[4] We can also adopt the notation gV for the covariant vector field (1-form) which maps W to $g(V, W)$, i.e., whose index expression is $g_{ab}V^b (= V_a)$. Similarly, if A is a covariant vector field, then $g^{-1}A$ has the index expression $g^{ab}A_b (= A^a)$. Thus we can write the above condition $gV = \lambda hV$, $h(V, V) = 1$.

Lorentz metric, namely, that it should admit some nowhere vanishing vector field (Euler characteristic vanishing; cf. Markus [13]).

1.6. *Remark.* If a space-time M is not time-orientable, there always exists a time-orientable space-time M' which is a twofold covering of M (see Markus [13]). This is not hard to see (for example, construct M' by choosing each point to represent a half-cone of timelike vectors at a point of M). The result of such a procedure is illustrated in Fig. 1.

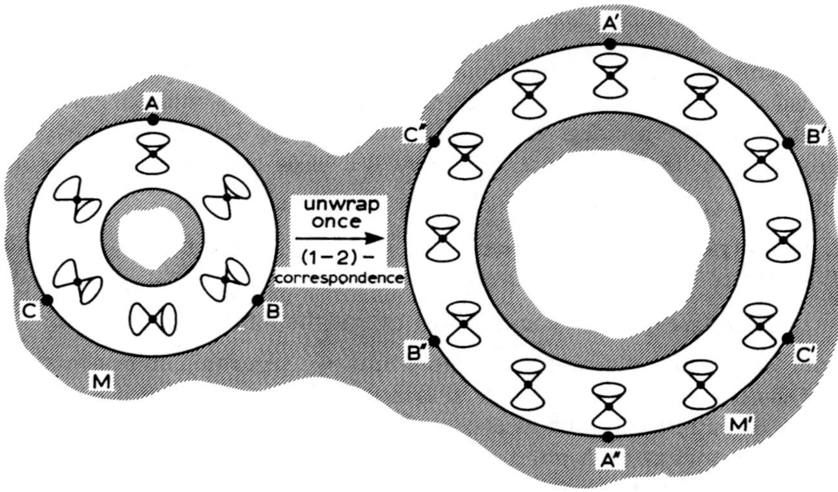

FIG. 1. *The space-time on the left is not time-orientable, but its twofold covering, depicted on the right, is. (The light cones are here drawn as three-dimensional, for descriptive purposes, even though the space-times are only two-dimensional; cf. also Fig. 21.)*

Many theorems about time-orientable space-times will be applicable also to non-time-orientable space-times, since the theorem may be referred to the time-orientable double coverings. In view of this fact, and also from the standpoints of "physical reasonableness" and mathematical convenience, I shall henceforth restrict all considerations to space-times which are *time-oriented.* (This restriction is often made in the definition of a space-time in any case.) The symbol M will, in fact, always denote a time-oriented space-time in these notes.

1.7. DEFINITION. A *path* is a continuous map $\mu:\Sigma \to M$, where Σ is a connected subset of $\mathbb{R}$ containing more than one point. This is a *smooth path* if μ is smooth with nonvanishing derivative $d\mu$ (the degree of smoothness being C^∞ unless otherwise stated). Thus, a path carries a parameter, the parameter range Σ being the path domain. ($\mathbb{R}$ denotes the field of real numbers.)

The term (smooth) *curve* will be used either for the image of such a map or (more correctly) for an equivalence class of paths equivalent under (smooth) parameter change (i.e., homeomorphisms or diffeomorphisms of the path domains); an *oriented* curve arises if the parameter change is required to be *monotonic.* A smooth path is called *timelike* if its tangent vector is timelike at every point; such a path is *future-oriented* if its tangent vector is future-pointing at every point.

We may also speak of smooth *causal* paths and future-oriented smooth causal paths, where the tangent vectors are allowed to be null as well as timelike. However, we shall be concerned later with such paths primarily when the path is *not* restricted to be smooth, in which case a somewhat different definition will be required.

A *timelike curve* is a curve defined by a smooth timelike path. The time-orientation of M assigns a canonical (future) orientation to any timelike curve, namely, that defined when the path is future-oriented. For this reason, and also owing to the fact that locally (and globally if M contains no closed timelike curves) the image in M of a smooth timelike path μ determines the equivalence class of μ under parameter change, it becomes generally unimportant to distinguish between the above alternative possibilities for the definition of a curve, when the curve is timelike. Thus, I shall use shorthand notations such as $\gamma \subset M$, when γ is a timelike curve, even when the equivalence class definition may be more appropriate.[5] (The same will apply to a causal curve.)

1.8. DEFINITION. To define an *endpoint* of a path μ, or of its associated curve, let Σ be the domain of μ and let $a = \inf \Sigma$, $b = \sup \Sigma$ (possibly $a = -\infty$ or $b = \infty$). Then $x \in M$ is an endpoint if for all sequences $\{u_i\} \in \Sigma$, $u_i \to a$ implies $\mu(u_i) \to x$ *or* $u_i \to b$ implies $\mu(u_i) \to x$. If μ is timelike (or causal) and future-oriented, then in the first case x is a *past* endpoint and in the second case a *future* endpoint.

For convenience I shall require all timelike or causal curves to *contain their endpoints*. (So, for a curve with two endpoints, Σ must be a closed interval.) This has the implication, for example, that the situations depicted in Fig. 2 are excluded;

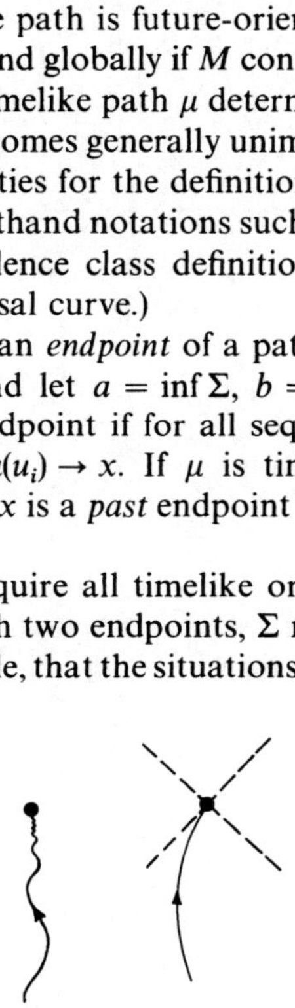

FIG. 2. *Such cases as these are to be excluded as smooth timelike curves because the future endpoints are required to be part of the curves. Our timelike curves must be smooth and timelike at their endpoints. (The convention employed here is the standard one that "time" proceeds from the bottom of the page to the top, with null lines depicted at 45°. Except when explicitly indicated otherwise, this standard convention will also be used in all other figures.)*

a timelike curve has to be smooth and strictly timelike *at* its endpoints. Hence it must be extendible as a timelike curve at any endpoint. A timelike curve (or path) without a future endpoint must extend indefinitely into the future; such a curve (or path) is called *future-endless*. Similarly a timelike curve (or path) without past

[5] The set inclusion symbol $\subset$ used here is reflexive, that is to say, $A \subset A$ is always valid. The boundary of a set A is denoted by ∂A, its closure by $\bar{A}$ and its complement in M by $\sim A$. The difference of two sets is denoted by $A - B\ (= A \cap (\sim B))$. The set of all x with property p is denoted by $\{x | p(x)\}$.

endpoint is called *past-endless*; if it has neither future nor past endpoint it is called, simply, *endless*.[6]

As a point set in M, a timelike curve will usually be a closed set (since any endpoints must be included), but not always, if no global causality restrictions on M are made, since situations such as that depicted in Fig. 3 may arise. Here a

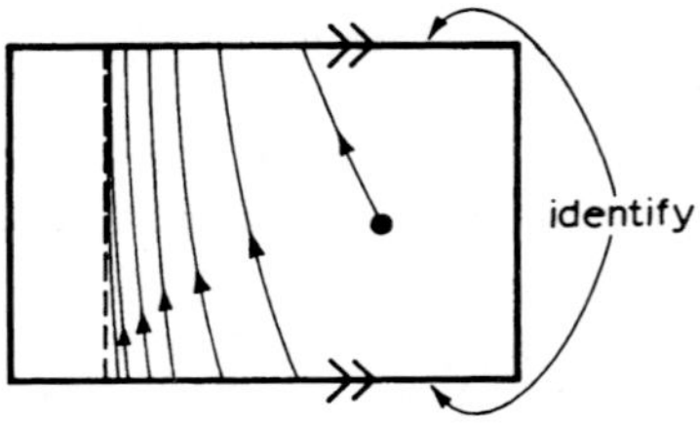

FIG. 3. *A timelike curve winds around endlessly within a compact set. This curve has no future endpoint and is an example of a future-endless timelike curve*

future-endless timelike curve is depicted which winds around endlessly within a compact region. (Other somewhat similar situations can also arise in which a timelike curve is not eventually contained within a compact region, but nevertheless keeps re-entering such a region.)

1.9. DEFINITION. The symbol ∇ will be used to denote the unique torsion-free connection on M under which g is covariantly constant (equivalently: under which the scalar product defined by g is preserved under parallel transport along any curve). An *affinely parameterized geodesic*, abbreviated *a. p. geodesic* is a path with tangent vector T satisfying $\nabla_T T = 0$ (i.e., $T^a \nabla_a T^b = 0$) at every point of the curve. The term *geodesic* will here refer to the curve associated with a path which is an a. p. geodesic. A geodesic is *timelike*, *null*, *spacelike* or *causal* according as T is timelike, null, spacelike, or either timelike or null. This holds at every point of the curve if it holds at any one of its points (trivially, since ∇ preserves scalar products between parallelly propagated vectors, and in particular it preserves $g(T, T) = T_a T^a$). A *degenerate geodesic* occurs when $T = 0$ (so the curve lies all at one point). Unless otherwise stated, all geodesics will be assumed to be *nondegenerate*. (In any case, a degenerate geodesic is not a smooth curve, according to 1.7, since $d\mu = 0$.)

1.10. DEFINITION. To proceed further we shall need some simple properties of the *exponential map*. For any $a \in M$, this is smooth (C^∞) map, denoted $\exp_a$ from some open subset of the tangent space $\underset{a}{T}$, into M. If $V \in \underset{a}{T}$, we define $\exp_a(V)$ to be the point p of M (if such exists) such that the affinely parameterized geodesic with tangent vector V at a and parameter value 0 at a acquires the parameter value 1 at p. If V has components (t, x, y, z) with respect to some basis for $\underset{a}{T}$, then t, x, y, z are called (Riemannian) *normal coordinates* of the point p. I shall often use such coordinates in the particular case when the null cone in $\underset{a}{T}$ is given by

[6] The term "inextendible" in place of "endless" has been used in some articles [6], [9].

$t^2 - x^2 - y^2 - z^2 = 0$ and $\partial/\partial t$ is future-pointing. These I call *Minkowski normal coordinates.*

The condition that $\exp_a$ map the whole of $\underset{a}{T}$ into M for every choice of $a \in M$ is that M be *geodesically complete*; that is to say, every affinely parameterized geodesic in M extends to arbitrarily large parameter values. But whether or not the whole of $\underset{a}{T}$ is mapped, it may well be that several different elements of $\underset{a}{T}$ are mapped to the same point of M, or that the map is badly behaved for certain elements of $\underset{a}{T}$ (because its Jacobian vanishes) so that the normal coordinate system breaks down at the corresponding point p of M. These situations are illustrated in Figs. 4, 5 and 6. We shall return to this question when we consider

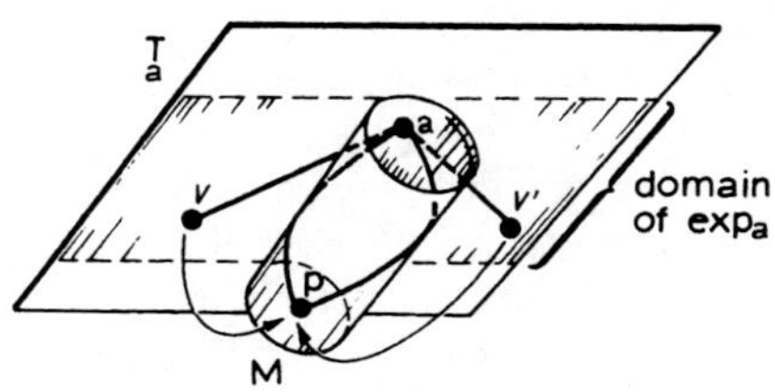

FIG. 4. *The Riemannian 2-space M is the surface of a finite cylinder. Here,* $\exp_a$ *maps a strip in* $\underset{a}{T}$ *onto M, wrapping it around M infinitely many times. The point $p \in M$ is the image of infinitely many points in* $\underset{a}{T}$*, in particular, of v and of v'*

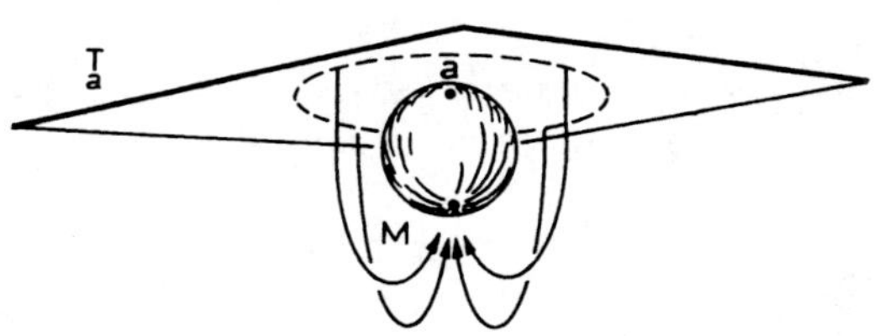

FIG. 5. *Here M is a unit sphere (a positive definite Riemannian 2-space). The circles of radius* π, 2π, 3π, $\cdots$ *with centers at the origin of* $\underset{a}{T}$ *each map to a single point on M under* $\exp_a$

conjugate points in 7.10. For the present we require the fact that for each $a \in M$ there is *some* star-shaped[7] neighborhood Q of the origin in $\underset{a}{T}$ such that $\exp_a$, restricted to Q, is a diffeomorphism (i.e., (t, x, y, z) form an allowable coordinate system for $\exp_a Q$). Then $\exp_a Q$ is called a *normal neighborhood* of a [14]. We can, in fact, always choose a normal neighborhood N of any $a \in M$ such that N is also a normal neighborhood of any *other* point $b \in N$. Such an N is called *simply convex*. The characteristic property [14] of a simply convex neighborhood N is that $N \subset M$ is open and that there is precisely one geodesic lying within N

[7] The rather nondescriptive term "star-shaped" simply means that if $V \in Q$, then $\lambda V \in Q$ for all $\lambda \in [0, 1]$; that is to say, all the rays through the origin are connected in Q.

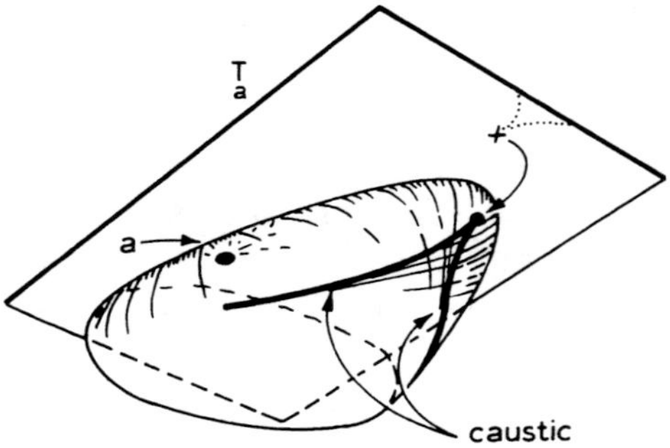

FIG. 6. *Again M is a positive definite Riemannian 2-space, but a little more general in shape than before. The map* $\exp_a$ *is still badly behaved in certain regions, having vanishing Jacobian on a curve on M referred to as a caustic. The caustic is the envelope of geodesics on M through a (i.e., roughly speaking, the locus of points where consecutive geodesics intersect)*

connecting each pair of points in N. It will be convenient to consider such sets frequently in this work. But since it will be convenient also to demand a few more properties, let me define a *simple region N* to be a simply convex open subset of the space-time M such that $\overline{N}$ is *compact* and is contained in a simply convex open set. Then we have the following properties (cf. [14], [35]).

1.11. PROPOSITION. *If N is a simple region, any two points p, q of* $\overline{N}$ *can be connected by a unique geodesic in N, denoted by pq. The geodesic pq is a continuous function of* $(p, q) \in \overline{N} \times \overline{N}$.

1.12. PROPOSITION. *The boundary* ∂N *of any simple region N is compact; any closed subset of N is compact.*

1.13. PROPOSITION. *The space-time M can be covered by a locally finite system of simple regions; any compact subset of M can be covered by a finite number of simple regions.*

1.14. *Remark.* Unlike the situation for positive definite Riemannian spaces, it is not true that every compact space-time is geodesically complete [15]. Nor is it necessarily true [16] for a geodesically complete space-time that $\exp_a$ maps to the whole of M. A counterexample is provided by anti-deSitter space (Fig. 7).

1.15. DEFINITION. Although the main applications will come considerably later (cf. 7.10) it will also be convenient at this point to digress a little and introduce the concept of a *Jacobi field* [14], [17], defined along an a. p. geodesic γ. A Jacobi field may be thought of as defining a "vector in the space of a. p. geodesics," that is to say (roughly speaking), it describes the relation between an a. p. geodesic and another one which lies infinitesimally close to it.

More precisely, let γ belong to a smooth 1-parameter system of a. p. geodesics. This can be described by a smooth map μ from a strip $\{(t, v) | t_0 < t < t_1, -\varepsilon < v < \varepsilon\}$ into M, where each path defined by setting $v =$ const. is an a. p. geodesic, parameterized by t, and γ is given by $v = 0$ (we can allow $-t_0$ or t_1 to be ∞ if necessary) (see Fig. 8). Denote the "coordinate vectors" on M by $T = \partial/\partial t$ and $V = \partial/\partial v$. (Strictly this should be $T = \mu_*(\partial/\partial t)$ and $V = \mu_*(\partial/\partial v)$, but I shall allow myself this sort of abuse of notation.) We can write (cf. 1.9): $T = T^a\nabla_a$,

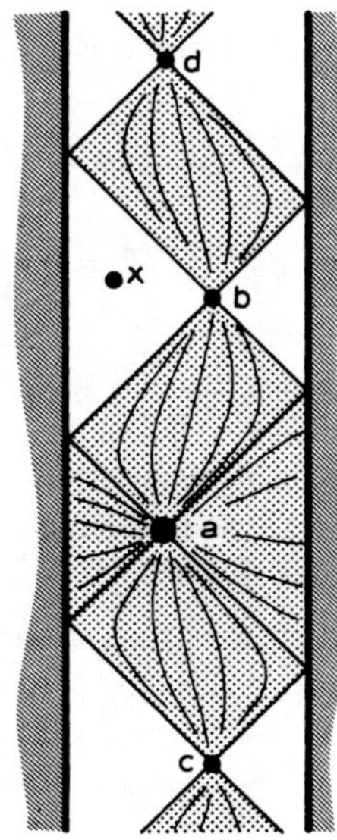

FIG. 7. *Anti-deSitter space (2-dimensional case). M is conformal to the strip* $-\pi/2 < x < \pi/2$ *of Minkowski 2-space, the metric of M being given by* $\sec^2 x(dt^2 - dx^2)$. *Alternatively M may be expressed as the universal covering space of the "sphere"* $T^2 + W^2 - X^2 = 1$ *in the Minkowski 3-space* $ds^2 = dT^2 + dW^2 - dX^2$ (*where* $\tan x = X$, $\tan t = T/W$).

All timelike geodesics through $a \in M$ *are focused again at the "antipodal" point b, and again at c, d, e,* $\cdots$. *Spacelike and null geodesics through a "go off to infinity" and never reach b, c, d,* $\cdots$, *so* $\exp_a$ *maps to the dotted region only, even though the space-time is geodesically complete* [16]. *The generalization to four dimensions is straightforward*

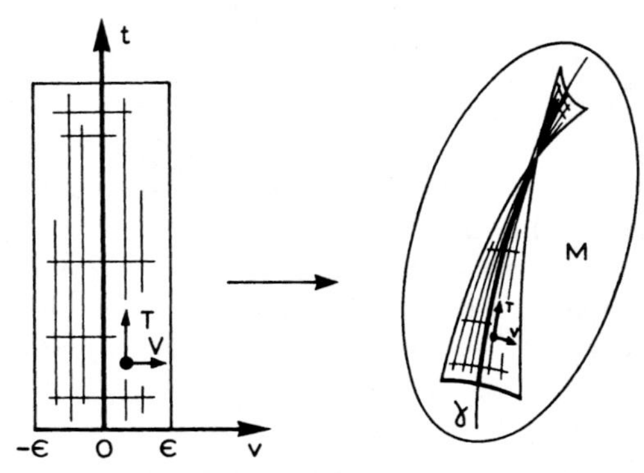

FIG. 8. *The map of the strip in the* (t, v)*-plane into M is smooth even though the image (and the inverse map) may be singular. The image of each line* $v =$ const. *is a geodesic in M (affinely parameterized by t). The vectors* $V = \partial/\partial v$ *define a Jacobi field along the geodesic* γ, *which "point" from* γ *to a neighboring geodesic* γ'

$V = V^a\nabla_a$, when the ∇_a operators act on scalars; otherwise, we write $T^a\nabla_a = \underset{T}{\nabla}$, $V^a\nabla_a = \underset{V}{\nabla}$. For convenience, set

$$D = T^a\nabla_a \quad (= T \text{ or } \underset{T}{\nabla}) \tag{1}$$

to denote propagation derivative along γ. Since T and V are coordinate vectors we have $[T, V] = 0$, so

(2) $$\underset{T}{\nabla} V = \underset{V}{\nabla} T, \quad \text{i.e., } DV^a = V^b \nabla_b T^a.$$

Since we have a. p. geodesics, we have $DT^a = 0$ ($\underset{T}{\nabla} T = 0$). Hence, differentiating (2) we arrive at:

(3) $$D^2 V^a = R^a{}_{bcd} T^b V^c T^d,$$

which is the familiar *Jacobi equation* (equivalently written $D^2 V = R(V, T)T$), or equation of *geodesic deviation.* (The Riemann tensor sign is here taken consistent with $(\nabla_a \nabla_b - \nabla_b \nabla_a)V^c = R^c{}_{dba} V^d$; equivalently $\{\underset{U}{\nabla}\underset{V}{\nabla} - \underset{V}{\nabla}\underset{U}{\nabla} - \underset{[U,V]}{\nabla}\}W = -R(U, V)W$.)

Any field of vectors V defined along γ and satisfying (3) is called a *Jacobi field.* Intuitively, the vectors V connect points of γ to corresponding points of some neighboring a. p. geodesic γ'. The solutions of (3) for V (given γ) form an 8-dimensional vector space ($2n$-dimensional, for an n-dimensional manifold), since (3) is linear in V. Any such solution is defined by knowledge of V and DV (which can be assigned arbitrarily) at any point of γ. This corresponds to the freedom in location or direction or scaling for a neighboring a. p. geodesic γ' to γ.

1.16. PROPOSITION. *If $T_a T^a$ is the same for each geodesic of the 1-parameter system, then $T_a V^a$ is constant along γ.*

Proof.

$$\begin{aligned} D(T_a V^a) &= T_a D V^a = T_a V^b \nabla_b T^a \\ &= \tfrac{1}{2} V^b \nabla_b (T_a T^a) = 0 \end{aligned}$$

(or we can write this $Tg(T, V) = g(T, \underset{T}{\nabla} V) = g(T, \underset{V}{\nabla} T) = \tfrac{1}{2}\underset{V}{\nabla} g(T, T) = 0$.)

1.17. *Remark.* If γ is timelike we can choose $T_a T^a = 1$ (if spacelike, $T_a T^a = -1$). With the scaling freedom for γ' eliminated, the allowable Jacobi fields along γ now form a 7-dimensional vector space. We can further insist that V is orthogonal to T all along γ (by 1.16) and we get a 6-dimensional space. (This is simply a question of suitably fixing the parameter origin for γ'.) A situation of this type arises if we consider timelike (or spacelike) a. p. geodesics with unit tangent vector, starting from a fixed p. This is because $V = 0$ at p, so certainly $T_a V^a = 0$. (It is important to note the fact that the map μ of the strip is still smooth at p, even though the image is a singular surface at p.)

If γ is null, the situation is slightly different. Here $T_a T^a = 0$, so neighboring *null* a. p. geodesics have 7 degrees of freedom. The scaling of the affine parameter on a null geodesic cannot be fixed in a natural way, so freedom in the parameter scaling still exists in addition to the freedom in origin of parameterization. The condition $T_a V^a = 0$ is now nothing to do with the origin of parameterization but states a geometrical relation between γ and γ' (namely, that they could be "neighboring generators of a null hypersurface"). As an example where $T_a V^a = 0$ is satisfied, consider null a. p. geodesics through a fixed point p. Then γ and γ' become neighboring generators of the light cone of p.

1.18. DEFINITION. If V is a Jacobi field defined on γ and V vanishes at two distinct points $p, q \in \gamma$, while not vanishing at all points of γ, then p and q are called a *pair of conjugate points* on γ. This concept will have great importance later (cf. 7.10). Roughly speaking, a pair of conjugate points occurs where two neighboring geodesics meet in two points. This arises when geodesics through p encounter a *caustic* at q, showing that we must expect the Riemannian normal coordinates defined by $\exp_p$ to break down as a coordinate system, at q, the Jacobian vanishing there (see Fig. 6). In fact the caustic could, in this context, be *defined* as the set of points of M conjugate to p on geodesics through p.

SECTION 2

Causality and Chronology

2.1. DEFINITION. A *trip* is a curve which is piecewise a future-oriented timelike geodesic. A trip *from x to y* is a trip with past endpoint x and future endpoint y. We write $x \ll y$ (read x *chronologically precedes* y) if and only if there exists a trip from x to y. Thus, the relation $x \ll y$ states the existence of points $x_0, x_1, \cdots, x_n$ with $n \geqq 1$, a timelike geodesic called a *segment* having past endpoint x_{i-1} and future endpoint x_i, for each $i = 1, \cdots, n$, where we set $x_0 = x$, $x_n = y$. Note that since the curves defined here are required to contain all their endpoints, the situation depicted in Fig. 9 (a "bad trip") in which the segments accumulate at a point p cannot occur.[1]

FIG. 9. *A "bad trip" has an infinite number of "joints" accumulating at p*

2.2. *Remark.* We shall see in 2.23 that timelike curves could equally well have been used in place of trips to define $\ll$, which would perhaps have been more physical, but trips turn out to be easier to handle mathematically. Compare [18].

Observe that in the above we could always choose $n = 1$ for $x \ll y$ in Minkowski space. On the other hand, space-times exist for which it is necessary for n to be allowed to be indefinitely large. An example (a "mutilated Minkowski space") is given in Fig. 10. A less artificial example, which shows that we need to allow $n \geqq 2$, is afforded by the anti-deSitter space (Fig. 7; see also Fig. 11).

2.3. DEFINITION. A *causal trip* is defined in the same way as a trip except that causal geodesics, *possibly degenerate*, replace the timelike geodesics of 2.1. We write $x \prec y$ (read x *causally precedes* y) if and only if there is a causal trip from x to y. See [18].

2.4. *Remark.* Note that $x \prec x$ for all $x \in M$, since degenerate causal geodesics are allowed. On the other hand, $x \ll x$ signifies the existence of a *closed trip* in M, that is, a trip whose past and future endpoints are identical. (Minkowski space, for example, possesses no closed trips.) A closed nondegenerate causal trip is signified by the existence of a pair of distinct points x, y such that $x \prec y$ and $y \prec x$.

[1] A trip with infinitely many segments is allowable of course, provided it is future- or past-endless.

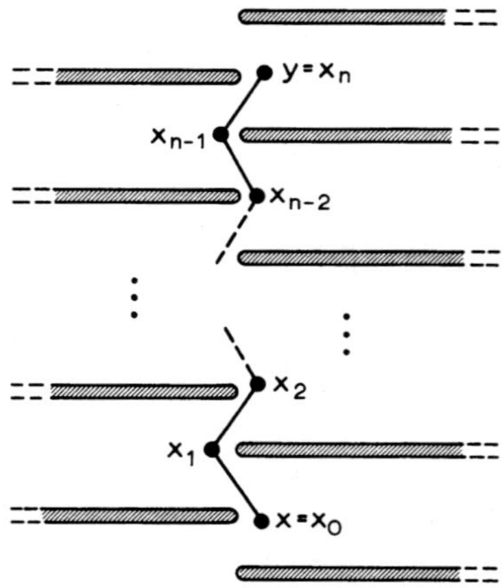

FIG. 10. *From Minkowski 2-space the half-lines $t = k$, $(-1)^k x \geqq 0$ are removed. To express the relation $x \ll y$, trips with arbitrarily large numbers of segments are required*

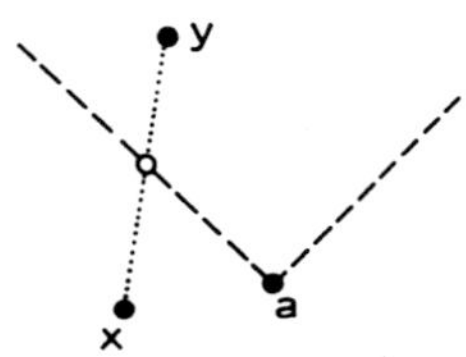

FIG. 11. *The space-time M is Minkowski space with one point removed. The set $J^+(a)$ is not closed since the null geodesic beyond the removed point, which extends that from a, is not part of $J^+(a)$, whereas it is part of $\partial J^+(a)$. (Small open circles in diagrams always denote removed points.)*

2.5. PROPOSITION.

$$a \ll b \quad \textit{implies} \quad a \prec b;$$

$$a \ll b, \quad b \ll c \quad \textit{implies} \quad a \ll c;$$

$$a \prec b, \quad b \prec c \quad \textit{implies} \quad a \prec c.$$

2.6. DEFINITION. The set $I^+(x) = \{y \in M | x \ll y\}$ is called the *chronological* (or open) *future* of x; $I^-(x) = \{y \in M | y \ll x\}$ is the *chronological past* of x; $J^+(x) = \{y \in M | x \prec y\}$ is the *causal future* of x; $J^-(x) = \{y \in M | y \prec x\}$ is the *causal past* of x. The chronological or causal future of a *set* $S \subset M$ is defined by $I^+[S] = \bigcup_{x \in S} I^+(x)$, $J^+[S] = \bigcup_{x \in S} J^+(x)$, respectively, and similarly for the pasts of S. (In general there will be a self-evident "duality" obtained by interchanging past and future in any result. The dual version of result will not normally be stated explicitly in what follows.) The slight abuse of notation $I^+[\gamma]$, etc., where γ is a trip, etc., will also be used.

2.7. *Remark.* In Minkowski space with the usual coordinates (t, x, y, z), if $a = (0, 0, 0, 0)$, then $I^+(a) = \{(t, x, y, z) | t > (x^2 + y^2 + z^2)^{1/2}\}$. Also $J^+(a)$ is the same but with "$\geqq$" replacing "$>$." Here $I^+(a)$ is an open set and $J^+(a)$ a closed set. In fact, every chronological future is open (cf. 2.9) but not all causal futures are closed. As an example of this, obtain the causal future $J^+(a)$ in Fig. 11.

2.8. PROPOSITION. *$I^+(a)$ is open for any $a \in M$.*

Proof. Let $x \in I^+(a)$; then there is a trip γ from a to x. Let $N \ni x$ be a simple region and let y be a point in N, other than x, on the terminal segment of γ. Now the vector $\exp_y^{-1}(x)$ is timelike and future-pointing (being a tangent to the terminal segment at y), and so belongs to the open set Q of timelike future-pointing vectors in $\exp_y^{-1}[N]$. Since $\exp_y$ is a homeomorphism in this neighborhood, it follows that $\exp_y Q$ is an open set in M (containing x) which lies in $I^+(y)$ and therefore in $I^+(a)$ (by 2.5), thus proving the result.

2.9. COROLLARY. *$I^+[S]$ is open, for any $S \subset M$.*

2.10. PROPOSITION. *$x \in I^+(y)$ if and only if $y \in I^-(x)$; $x \in J^+(y)$ if and only if $y \in J^-(x)$.*

2.11. PROPOSITION. *$I^+[S] = I^+[\bar{S}]$.*

Proof. If $y \gg x$, $x \in \bar{S}$, then $y \gg z$, $z \in S$ since $I^-(y)$ is open.

2.12. PROPOSITION. *$I^+[S] = I^+[I^+[S]] \subset J^+[S] = J^+[J^+[S]]$.*

Proof. This follows from 2.5, from the fact that $a \ll b$ implies the existence of c with $a \ll c \ll b$ and from the corresponding statement for $a \prec b$.

2.13. DEFINITION. Let N be a simple region and define [36], [19] the *world-function* $\Phi: N \times N \to \mathbb{R}$ by $\Phi(x, y) = g(\exp_x^{-1}(y), \exp_x^{-1}(y))$; in other words, $\Phi(x, y)$ is the squared length of the geodesic xy. Clearly $\Phi(x, y) = \Phi(y, x)$ and is positive, negative or zero according as xy is timelike, spacelike or null.

2.14. PROPOSITION. *$\Phi(x, y)$ is a continuous function of (x, y) in $N \times N$.*

Proof. See 1.11, [36], [19].

2.15. LEMMA. *The point $p \in N$ being kept fixed, the hypersurfaces $H_{p,K} = \{x | \Phi(p, x) = K\}$ are smooth in N (except at $x = p$) and are spacelike, timelike or null according as the constant K is positive, negative or zero. Furthermore, the geodesic px is normal to $H_{p,K}$ at x.*

Proof. The smoothness follows from the fact that $\exp_p$ is well-behaved in N, the equation of $H_{p,K}$ in Minkowski normal coordinates being $t^2 - x^2 - y^2 - z^2 = K$, which is smooth (except at the origin, when $K = 0$). A smooth hypersurface is said to be spacelike, timelike, or null according as its normal vectors are timelike, spacelike, or null. Let q be a point of $H_{p,K}$ and V a tangent vector to $H_{p,K}$ at q. Allowing q to vary on $H_{p,K}$ along a curve with tangent vector V, so that pq describes a 1-parameter system of a. p. geodesics of squared length K, we see that V belongs to a Jacobi field vanishing at p. Hence, by 1.16, V must be orthogonal, at q, to the direction of pq. The result follows.

2.16. LEMMA. *Let N be a simple region. Suppose $a, b, c \in \bar{N}$ are such that ab and bc are both future-causal, having distinct directions at b if both are null, or suppose a timelike curve or trip γ exists in $\bar{N}$ from a to c. Then ac is future-timelike.*

Proof. Consider $\Phi(x) = \Phi(a, x)$, as x varies from a to c along $\beta = ab \cup bc$ or along γ. As x proceeds in a future-causal direction defined by the vector T, the rate of change of Φ is measured by $T^i\nabla_i\Phi\,(= T(\Phi) = d\Phi(T) = g(g^{-1}\,d\Phi, T)) = g_{ij}T^i\nabla^j\Phi$. This, by 2.15, is nonnegative whenever ax is future-causal ($\nabla^i\Phi$, or $g^{-1}\,d\Phi$ being normal to $\Phi = \text{const.}$, i.e., to $H_{p,\Phi}$) and strictly positive unless ax

is null and T tangent to ax. (The scalar product of two future-causal vectors is nonnegative, being zero only if both are null and proportional.) Hence $\Phi(c) = \Phi(a, c) > 0$ and ac must be *future*-timelike, since $\exp_a^{-1} x$ never leaves the future component of the timelike vectors at a.

2.17. *Remark.* The proof of the lemma in 2.16 is based on 2.15 for causal ax ($K \geqq 0$). Alternatively, the argument could equally well have been given using the result only for *null* ax ($K = 0$). Essentially we require only the fact that the light cone $H_{a,0}$, being a null hypersurface (except at a) cannot be crossed from the inside to the outside by β or γ.

It is of some interest to note that the lemma in 2.16 is false for a $\tilde{\nabla}$ with torsion (but with $\tilde{\nabla}g = 0$ still holding). This is illustrated in Fig. 12. The light cone with respect to $\tilde{\nabla}$ is a *timelike* surface, being generated by null curves which are geodesics with respect to $\tilde{\nabla}$, but curl into the inside of the light cone with respect to ∇. Thus, β or γ can escape from inside to outside the light cone.

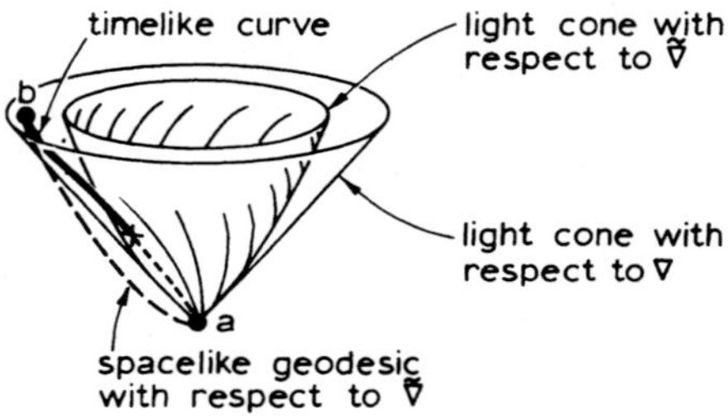

FIG. 12. *If we replace the Riemannian connection ∇ by another connection $\tilde{\nabla}$ which still preserves the metric ($\tilde{\nabla}g = 0$) but which possesses torsion, then 2.16 becomes untrue. We have a timelike curve connecting a to b, but the geodesic ab (according to $\tilde{\nabla}$) is spacelike*

2.18. PROPOSITION.

$$a \ll b, \quad b \prec c \quad \textit{implies} \quad a \ll c;$$

$$a \prec b, \quad b \ll c \quad \textit{implies} \quad a \ll c.$$

Proof. Without loss of generality, suppose $a \ll b$ and $b \prec c$. Let α be a trip from a to b and γ a causal trip from b to c. Then γ (being compact)[2] can be covered by a finite number of simple regions $N_1, \cdots, N_r$. (It is clear that we can assume that γ has no closed-loop parts, since redundant portions can be deleted.) Set $x_0 = b \in N_{i_0}$, say. Let x_1 be the future endpoint of the connected component of $\gamma \cap \bar{N}_{i_0}$ from x_0. Choose $y_1 \in N_{i_0}$ on the final segment of α, with $y_1 \neq x_0$ (see Fig. 13). Then by the lemma in 2.16, $y_1 x_1$ is future-timelike. Now, either $x_1 = c$, in which case the result is established, or $x_1 \notin N_{i_0}$, whence $x_1 \in N_{i_1}$, say. In the latter case, let x_2 be the future endpoint of the connected component of $\gamma \cap \bar{N}_{i_1}$ from x_1 and choose $y_2 \in N_{i_1}$ on $y_1 x_1$ with $y_2 \neq x_1$. Then either $x_2 = c$, in which case we are finished, or we can repeat the argument. The process must eventually terminate, since there are a finite number of connected components of the $\gamma \cap \bar{N}_i$.

[2] Cf. 1.13.

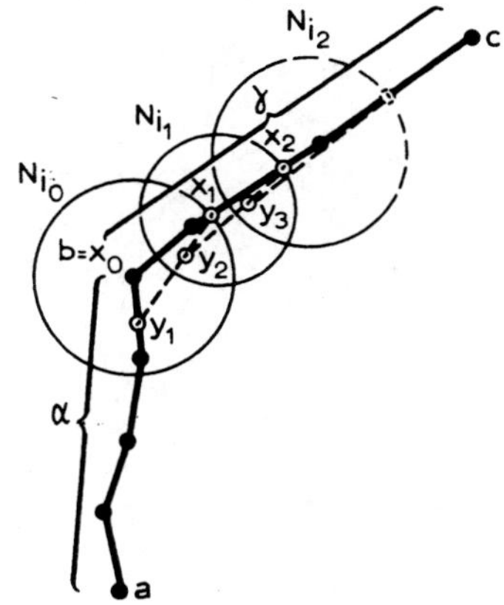

FIG. 13. *Construction for 2.18, to show that the trip α from a to b together with the causal trip γ from b to c can be replaced by a single trip from a to c*

2.19. PROPOSITION. *If α is a null geodesic from a to b, and β is a null geodesic from b to c, then either $a \ll c$ or else $\alpha \cup \beta$ constitutes a single null geodesic from a to c.*

Proof. If $\alpha \cup \beta$ fails to constitute a single geodesic, this is because the future direction of α at b does not agree with that of β at b (a "joint"). By 2.16, if x on α and y on β are sufficiently close to (but distinct from) b, then there is a timelike geodesic from x to y. Thus $a \prec x \ll y \prec c$, whence $a \ll c$ by 2.18.

2.20. PROPOSITION. *If $a \prec b$ but $a \not\ll b$, then there is a null geodesic from a to b.*

Proof. Let γ be a causal trip from a to b. If γ contains a timelike segment, then repeated application of 2.18 yields $a \ll b$. If all segments of γ are null, then repeated application of 2.19 yields $a \ll b$ unless γ is a null geodesic.

2.21. *Remark.* The relation: "$a \prec b$ but $a \not\ll b$"; is sometimes written $a \to b$ (or $a \nearrow b$) and is termed *horismos* [18], but I shall not concern myself with it explicitly here. The concepts of $\prec$, $\ll$ and $\to$ can refer to sets M more general than space-times, e.g., to a *causal space* (see Kronheimer and Penrose [18]), defined by relations $\prec$, $\ll$ on a set M subject to 2.5 and 2.18, and, in addition, to the requirements that $a \ll a$ hold for no a and that $a \prec b$, $b \prec a$ hold for no distinct pair a, b (stating the exclusion of "closed trips" or "closed causal trips").

2.22. *Remark.* The converse of 2.20 is false. (In the example illustrated in Fig. 14, there is a null geodesic from a to b, but $a \ll b$.) Observe, also, that we can have two distinct null geodesics from a to c and *not* have $a \ll c$ (cf. Fig. 14), but it is a consequence of 2.19 that any point x on the continuation of either geodesic beyond c must satisfy $a \ll x$.

2.23. PROPOSITION. *$a \ll b$ if and only if there is a timelike curve γ from a to b.*

Proof. Suppose γ exists. Cover γ with a finite number of simple regions N_i. Let $x_0 = a \in N_{i_0}$ and let x_1 be the future endpoint of the connected component of $\gamma \cap \bar{N}_{i_0}$, from x_0. Then by 2.16, x_0x_1 is future-timelike. Either $x_1 = b$, in which case $a \ll b$ as required, or else $x_1 \notin N_{i_0}$ so $x_1 \in N_{i_1}$, say. Let x_2 be the future endpoint of the connected component of $\gamma \cap \bar{N}_{i_1}$, from x_1. Then x_1x_2 is future-

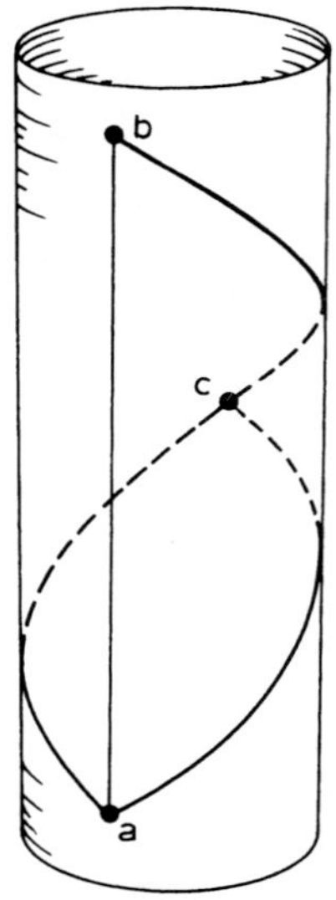

FIG. 14. *A two-dimensional "Einstein Universe" constructed by identifying* $(-1, t)$ *with* $(1, t)$ *in the strip* $-1 \leqq t \leqq 1$ *of Minkowski 2-space. Here* a *is* $(0, 0)$, b *is* $(0, 2)$ *and* c *is* $(1, 1)$. *We have* $a \ll b$ *even though* b *lies on a null geodesic through* a

timelike. Either $x_2 = b$, whence $a \ll b$, or else $x_2 \notin N_{i_1}$ so $x_2 \in N_{i_2}$ and the argument can be repeated. This terminates since there are a finite number of connected components of the $\gamma \cap \overline{N}_i$.

Conversely suppose $a \ll b$ and let α be a trip from a to b. I shall show that the "joints" of α can be smoothed so as to yield a timelike curve. Let μ and λ be consecutive segments of α. Let q be a point which is the future endpoint of the timelike geodesic λ and the past endpoint of the timelike geodesic μ. Consider $\exp_q^{-1}$ in some simple region $N \ni q$ and choose standard Minkowski coordinates (t, x, y, z) in $\underset{q}{T}$ so that the points of $\exp_q^{-1} \mu$ and $\exp_q^{-1} \lambda$ have coordinates of the form $(\tau, \tau \tan \chi, 0, 0)$ and $(-\tau, \tau \tan \chi, 0, 0)$, respectively, where τ varies over nonnegative values and where χ is fixed and satisfies $0 \leqq \chi < \pi/4$. Choosing $\tau_0 > 0$, connect $(-\tau_0, \tau_0 \tan \chi, 0, 0)$ to $(\tau_0, \tau_0 \tan \chi, 0, 0)$ by a C^∞ curve η in $\underset{q}{T}$ which joins on to $\exp_q^{-1} \lambda$ and $\exp_q^{-1} \mu$ smoothly (C^∞) and which is everywhere timelike according to the Minkowski metric $(dt^2 - dx^2 - dy^2 - dz^2)$ in $\underset{q}{T}$. For example, we could take η to be given by

$$R \cos\left(\frac{\theta\pi}{\pi - 2\chi}\right) = \exp\left(R^2 \sin^2\left(\frac{\theta\pi}{\pi - 2\chi}\right) - 1\right)^{-1},$$

where $t = \tau_0 R \sin \theta$, $x = \tau_0 R \cos \theta$ and $|R| \leqq 1$, $|\theta| \leqq \pi/2$. Measuring "angles" according to a "standard Euclidean metric" $dt^2 + dx^2 + dy^2 + dz^2$, we see that the slope of η is bounded away from the null cone in $\underset{q}{T}$, by an angle $\varepsilon\,(>0)$, say, where ε depends on χ but need not depend on τ_0. By choosing a small enough neighborhood of q in M we can ensure that the "error" in the slopes of the images of the null cones in M under $\exp_q^{-1}$ is less than ε. Hence, choosing τ_0 small enough, we ensure that $\exp_q \eta$ is timelike in M, thus achieving the required smoothing of the "joint" in $\lambda \cup \mu$.

2.24. *Remark.* Although 2.23 has some intrinsic interest in showing that trips and timelike curves are equivalent for defining the relation $\ll$, it will not in fact be required for any of the later results. All arguments can be carried out directly in terms of trips without any mention of smooth timelike curves.[3] On the other hand, the systematic use of timelike curves would be a little more awkward to handle since "smoothing arguments" would be required at various places (cf. 2.18 for example).

There is a similar result to 2.23 for causal trips (trivially, since by 2.20 and 2.23 a null geodesic or a timelike curve connects any two points for which $a \prec b$). However, I shall not restrict myself just to *smooth* causal curves here, since the role of a causal curve will be as a *limit* of timelike curves (or trips). A limit of a sequence of smooth curves need not be smooth. Let us therefore make the following definition which admits, under the term "causal curve," all such appropriate limits (cf. [21], [22], [6]).

2.25. DEFINITION. A curve γ is a *causal curve* if and only if for all $a, b \in \gamma$ and for every open set Q containing the portion[4] of γ from a to b, there is a causal trip from a to b (or from b to a) lying entirely in Q.

2.26. *Remark.* Although a causal curve γ need not be smooth, there is a restriction on its "degree of wildness" imposed by the fact that it satisfies a Lipschitzian type of condition. As a consequence, γ must possess a tangent almost everywhere (remark due to R. P. Geroch), even though examples can be concocted in which γ fails to have a tangent at a set of points dense on γ.

[3] Except, strictly speaking, that given for 8.8.

[4] If the reader is concerned about a slight illogicality here, in the confusion of two notions of "curve," he may care to rephrase the statement (i.e., "the portion of γ from a to b" refers to the equivalence class of paths under parameter change, whereas to be contained in Q it must be a point set). This kind of looseness of terminology is also to be found in many other places in these notes.

SECTION 3

Properties of Pasts and Futures

3.1. DEFINITION. A set $F \subset M$ is called a *future set* if $F = I^+[S]$ for some $S \subset M$ (cf. 2.6). Clearly F is a future set if and only if $F = I^+[F]$ (cf. 2.12). A future set F therefore has the property (shared by certain other sets, cf. 3.5) that: if $x \in F$ and $x \ll y$, then $y \in F$. Any future set is open, by 2.9.

Similarly P is called a *past set* if $P = I^-[S]$ for some $S \subset M$; equivalently, if $P = I^-[P]$. Any past set is likewise open. Many of the results which follow will have counterparts ("duals") for which "past" and "future" are interchanged. These dual results will be taken as understood and not normally stated explicitly.

3.2. *Side remark.*[1] A past set which is not the union of two past sets unless one contains the other, is called an *irreducible past* (abbreviated IP). A past set P is an IP if and only if it is of the form $P = I^-[\gamma]$, where γ is a trip (or timelike or causal curve). Any set $I^-(p)$ with $p \in M$ is an IP. An IP which is not of this form is called a terminal IP (abbreviated TIP). Any TIP has the form $I^-[\gamma]$, where γ is a future-endless trip. IF's and TIF's can be defined dually. Provided M satisfies suitable causality requirements (cf. 4.2), the TIP's and TIF's provide a convenient means of defining boundary points ("points at infinity" or "singularities") for a space-time M. These matters will not be entered into here, however.

3.3. PROPOSITION. *If F is a future set, $\bar{F} = \{x | I^+(x) \subset F\}$.*

Proof. Suppose $I^+(x) \subset F$. Then any trip from x contains points arbitrarily close to x lying in F, so $x \in \bar{F}$. Conversely, suppose $x \in \bar{F}$. Let $y \in I^+(x)$, so $x \in I^+(y)$. But $I^-(y)$ is open, so it contains some point $z \in F$. Thus $z \ll y$, implying $y \in F$ as required.

3.4. PROPOSITION. *Let F be a future set. Then:*

$$\bar{F} = \sim I^-[\sim F],$$

$$\partial F = \{x | I^+(x) \subset F \text{ and } x \notin F\}$$

$$= (\sim F) \cap (\sim I^-[\sim F]),$$

$$F = I^+[\bar{F}].$$

Proof. Exercise.

[1] See [23] for a full discussion.

3.5. PROPOSITION. *Let $Q \subset M$. Then the following are equivalent:*

$$I^+[Q] \subset Q,$$

$$I^-[\sim Q] \subset \sim Q,$$

$$I^+[Q] \cap I^-[\sim Q] = \varnothing,$$

$$\operatorname{int} Q = I^+[Q],$$

$$\partial Q = (\sim I^+[Q]) \cap (\sim I^-[\sim Q]).$$

Proof. Exercise. The proofs are facilitated if we bear in mind that the conditions on Q negate the possibility of having $a \in Q$, $b \in \sim Q$ and $a \ll b$.

3.6. PROPOSITION. *If $I^+[Q] \subset Q$ and Q is open, then Q is a future set.*

Proof. The result is immediate from 3.5: $Q = \operatorname{int} Q = I^+[Q]$.

3.7. PROPOSITION. *The union of any system of future sets is a future set; the intersection of two future sets is a future set.*

Proof. Clearly $\bigcup_i I^+[S_i] = I^+[\bigcup_i S_i]$, where i indexes the system. For the second part, observe that $I^+[Q] \subset Q$ and $I^+[R] \subset R$ together imply $I^+[Q \cap R] \subset Q \cap R$. The result then follows from 3.6.

3.8. PROPOSITION. *If $p \prec q$, then $I^+(p) \supset I^+(q)$.*

Proof. Immediate from 2.18.

3.9. PROPOSITION. $J^+[S] \subset \overline{I^+[S]}$.

Proof. If $y \in J^+[S]$, then $x \prec y$ for some $x \in S$, so the result follows from 3.8 and 3.3.

3.10. *Remark.* The converse of the proposition in 3.8 is false in many space-times (e.g., Minkowski space with the origin removed, where p and q have co-ordinates $(-1, -1, 0, 0)$ and $(1, 1, 0, 0)$ respectively), but it turns out to be true if M is any globally hyperbolic space-time (cf. 5.24), e.g., Minkowski space. Furthermore, space-times exist for which $I^+(p) \supset I^+(q)$ but not $I^-(p) \subset I^-(q)$ (e.g., p, q as above and M as Minkowski space less the half-plane $t \leqq 0$, $x = 0$).

3.11. DEFINITION. A set $S \subset M$ is called *achronal* if no two points of S are chronologically related (i.e., if $x, y \in S$, then $x \not\ll y$). (The term "semispacelike" has been used for the same concept [8], [9].)

3.12. *Remark.* A set can be locally spacelike without being achronal. Various examples are indicated in Figs. 15, 16 and 17. An achronal set can be null and

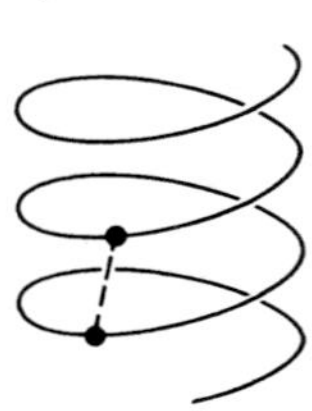

FIG. 15. *This curve in Minkowski space is locally spacelike, but it is not an achronal set since it contains pairs of points with a timelike separation. (This example incidentally shows that no useful concept of spacelike separation between points can be obtained from the condition that a spacelike curve connects them. Any pair of points in any space-time, of more than two dimensions, can be connected by a smooth spacelike curve.)*

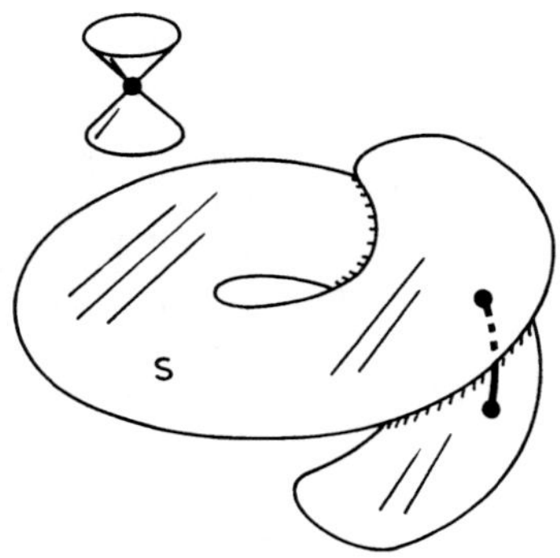

FIG. 16. *Even a smooth spacelike hypersurface in Minkowski space need not be achronal (although it would need to possess an "edge")*

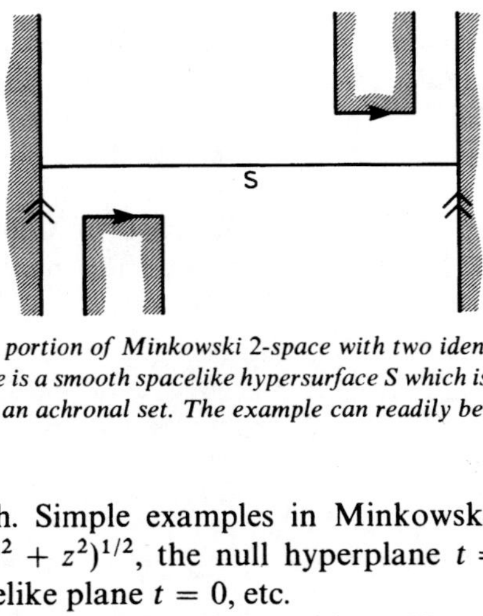

FIG. 17. *In this space-time (a portion of Minkowski 2-space with two identifications) there are no closed causal trips, but there is a smooth spacelike hypersurface S which is compact (and with no "edge") but which is still not an achronal set. The example can readily be made four-dimensional. It has relevance to* 8.3

it need not be smooth. Simple examples in Minkowski space are: the future light cone $t = (x^2 + y^2 + z^2)^{1/2}$, the null hyperplane $t = z$, the null line $t - z = x = y = 0$, the spacelike plane $t = 0$, etc.

3.13. DEFINITION. A set $B \subset M$ is called an *achronal boundary* if it is the boundary of a future set, i.e., $B = \partial I^+[S]$. Clearly no two points on the boundary of a future set F can be chronologically related ($I^+[\bar{F}] \cap \partial F = \varnothing$).[2] Thus any achronal boundary must in fact be an achronal set. (The term "semispacelike boundary" or "SSB" has sometimes been used in place of "achronal boundary"; cf. [8], [9].) The concept is actually time-symmetric as follows from the next proposition.

3.14. PROPOSITION. *B is an achronal boundary if and only if $B = \partial I^-[T]$ for some $T \subset M$.*

Proof. Suppose B is an achronal boundary, i.e., $B = \partial F$, where F is some future set. Then $T = {\sim}F$ will do. This follows because $I^-(x) \subset I^-[{\sim}F]$ if and only if $x \notin F$, and $x \notin I^-[{\sim}F]$ if and only if $I^+(x) \subset F$ (cf. 3.4). The converse in 3.14 is just the time-reverse of this.

3.15. PROPOSITION. *If B ($\neq \varnothing$) is an achronal boundary, then there is a unique future set F and a unique past set P such that F, P and B are disjoint with M*

[2] By 3.4.

$= P \cup B \cup F$. Then $B = \partial F = \partial P$. Furthermore, any trip or timelike curve from a point of P to a point of F must meet B in a unique point.

Proof. By 3.13 and the construction in 3.14, a future set F and past set P exist satisfying $B = \partial F = \partial P$, where $P = I^-[\sim F] = \sim(F \cup B)$ (cf. 3.4), $B \cap F = \emptyset$. Thus M is the union of the disjoint sets P, F and B as required. Before establishing uniqueness, let us examine the final part. Assume M is the union of disjoint sets B, $P = I^-[P]$ and $F = I^+[F]$. Let γ be a trip or timelike curve from $a \in P$ to $b \in F$. The sets $\gamma \cap (\sim F)$ and $\gamma \cap (\sim P)$ are both closed and together they exhaust γ. Therefore they are not disjoint, so γ meets B in a point, unique since B is achronal.

Suppose, now that M is also the union of disjoint sets B, $P' = I^-[P']$, $F' = I^+[F']$. If this decomposition is distinct from the earlier one, either $P \cap F'$ or $F \cap P'$ must be nonempty. Suppose $x \in P \cap F'$ (the case $x \in F \cap P'$ is exactly similar) and that $y \in B$. Since M is connected (and therefore arcwise connected), a curve ζ exists on M connecting x to y. It is clear that ζ can be taken to consist of a sequence of trips zig-zagging (if necessary) backwards and forwards in time. Buy ζ cannot leave P without crossing B in a future direction and therefore *entering* F'. Similarly, ζ cannot leave F' without crossing B in a past direction, therefore entering P. Thus ζ must remain in $P \cap F'$. But we have $y \notin P \cap F'$. This contradiction establishes the result.

3.16. *Remark.* In Minkowski space it always turns out that the F and P of 3.15 are given by $F = I^+[B]$, $P = I^-[B]$. However, for many space-times this need not hold. See, for example, the space-time illustrated in Fig. 18, which consists of

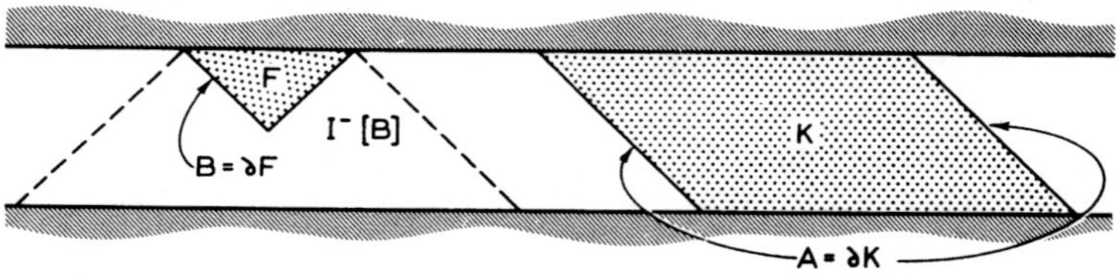

FIG. 18. *This M consists of that part of Minkowski space for which $0 < t < 1$. Here B is an achronal boundary for which $P \neq I^-[B]$ (P as in 3.15); that is to say, $B \neq \partial I^-[B]$, and for which B is a proper subset of another achronal boundary, namely $\partial I^-[B]$. Also, A is an achronal set which is a boundary, but A is not an achronal boundary*

a "horizontal" strip of Minkowski space. This example also shows that not every achronal boundary is a maximal achronal set. It also illustrates another pertinent fact: not every achronal set, which is the boundary of some other set, need be an "achronal boundary" as defined above. The trouble, here, arises from the fact that the two parts of the achronal set $A = \partial K$ have, in an appropriate sense, opposite orientation with respect to future directions. If the orientations are properly taken into account, then it is not hard to establish that every achronal set which is properly the boundary of another set is indeed an achronal boundary.

Achronal boundaries need not be smooth. Nevertheless they are quite "reasonable" sets as the next result shows.

3.17. LEMMA. *Any achronal boundary B is a topological (i.e., C^0) 3-manifold (that is, a continuous imbedded hypersurface).*

Proof. We have to establish that B is locally homeomorphic to E^3. Let P and F be as in 3.15 and choose $a \in B$. Let $N \ni a$ be a simple region and consider $\exp_a$ in N. Choose standard Minkowski coordinates in $\underset{a}{T}$ as normal coordinates for N. Let $Q \subset N$ be a region defined by $|t| \leqq \rho$, $x^2 + y^2 + z^2 < \frac{1}{4}\rho^2$ for some suitable $\rho > 0$. Choose ρ sufficiently small that curves in $\exp_a^{-1}[Q]$, which are "causal" with respect to the modified Minkowski metric $\frac{1}{4}dt^2 - dx^2 - dy^2 - dz^2$, map under $\exp_a$ to timelike curves in Q. In particular, $\exp_a$ will map each coordinate line $x, y, z =$ const. to a timelike curve $\eta_{x,y,z}$ in Q. Now the points in Q with normal coordinates $(-\rho, x, y, z)$ and (ρ, x, y, z) must lie in $I^-(a)$ and $I^+(a)$, respectively (since the geodesics connecting them to a are timelike). Hence, they must lie in P and F, respectively. Thus by 3.15, $\eta_{x,y,z}$ meets B in a unique point $b(x, y, z)$. This establishes a one-to-one mapping (an injection) between $B \cap Q$ and the interior of a sphere of radius $\frac{1}{2}\rho$ in $\mathbb{R}^3$. It remains to show that this mapping is continuous. But this follows because if $b(x, y, z)$ and $b(x + x_0, y + y_0, z + z_0)$ have a t-coordinate differing by more than $2(x_0^2 + y_0^2 + z_0^2)^{1/2}$, then these points must be chronologically related, the relevant curve described by $(t \pm 2\varepsilon(x_0^2 + y_0^2 + z_0^2)^{1/2}, x + \varepsilon x_0, y + \varepsilon y_0, z + \varepsilon z_0)$ as ε varies from 0 to 1 being necessarily timelike (since it is causal with respect to $\frac{1}{4}dt^2 - dx^2 - dy^2 - dz^2$). Since B is achronal, the t-coordinate difference must thus tend to zero as $(x_0, y_0, z_0) \to (0, 0, 0)$.

3.18. *Remark.* Certain types of achronal boundary of particular interest turn out to be (in a certain well-defined sense) null hypersurfaces. Let me illustrate the situation with a few examples. Set $B = \partial I^+[S]$ and take M to be Minkowski space. If $S = \{a\}$ for some $a \in M$, then B is the light cone of a and is a smooth null hypersurface except at a. If $S = \gamma$, where γ is the timelike curve $x = (1 + t^2)^{1/2}$, $y = z = 0$, then B is the hyperplane $t + x = 0$ which is smooth and null everywhere (see Fig. 19). Finally, if S is the spacelike 2-sphere $t = 0 = x^2 + y^2 + z^2 - 1$, the hypersurface B fails to be smooth and null on S, and also at the point p with

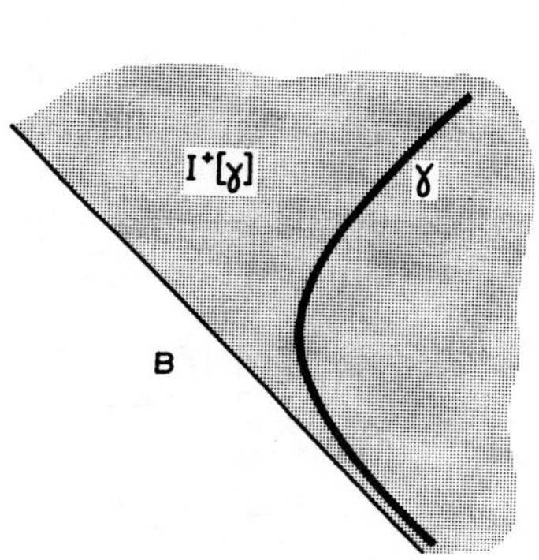

FIG. 19. *The curve γ is the world-line of a uniformly accelerating particle in Minkowski space. Then $B = \partial I^+[\gamma]$ is smooth and null everywhere (being a null hyperplane)*

coordinates (1, 0, 0, 0). (See Fig. 20 for the analogue of this when M is 3-dimensional Minkowski space.) Notice, however, that in this example (as in the previous ones) every point q of B which is not on S $(=\bar{S})$, including $q = p$, has the property that some null geodesic on B has q as its future endpoint. This property is, in fact, quite general, and is a consequence of the next lemma and its corollary.

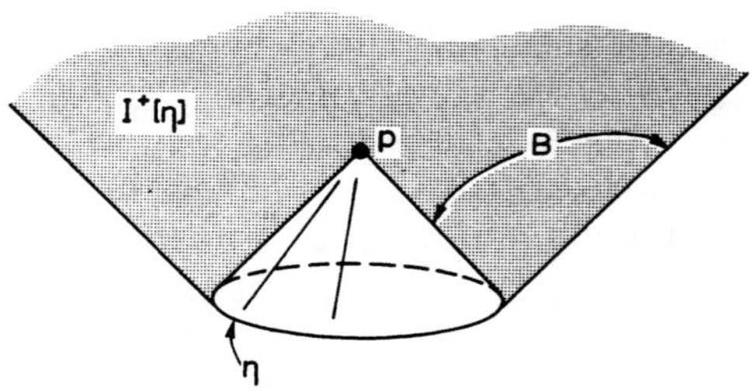

FIG. 20. *In three-dimensional Minkowski space, the null hypersurface $B = \partial I^+[\eta]$ is singular at p, this point lying away from the spacelike circle η given by $t = 0 = x^2 + y^2 - 1$, at which B is also singular as a hypersurface*

3.19. LEMMA. *Let F be a future set with $B = \partial F$. Let $x \in B$ and suppose an open set $Q \ni x$ exists such that:*

(a) *for any $y \in Q \cap F$ there is a trip γ from a point $z \in F - Q$ to y;*

or, equivalently:

(b) $F = I^+[F - Q]$.

Then B contains a null geodesic with future endpoint x.

Proof. Let us first establish that (a) and (b) are in fact equivalent. That (b) implies (a) is obvious. Conversely, suppose (a) holds. We must show that $y \in F$ implies the existence of $z \in F - Q$ such that $z \ll y$. But since F is a future set there certainly exists $w \in F$ with $w \ll y$. If $w \in F - Q$, we take $z = w$; if $w \in F \cap Q$, we invoke (a) to obtain $z \in F - Q$ and $z \ll w \ll y$. Thus (b) holds.

To establish the lemma, let $N \subset Q$ be a simple region containing x. Let $y_1, y_2, \cdots \in N \cap F$ be a sequence converging on x. (Clearly $x \in \overline{N \cap F}$, since $x \in \bar{F}$ and $N \ni x$ is open.) Each y_i is the future endpoint of a trip γ_i from some $z_i \in F - Q \subset F - N$. Let $v_i \in F \cap \partial N$ be the past endpoint of the connected component of $\gamma_i \cap \bar{N}$ which terminates at y_i. By 2.16, the geodesic $v_i y_i$ is timelike. Since ∂N is compact (cf. 1.12), an accumulation point v of the $\{v_i\}$ must exist, with $v \in \bar{F} \cap \partial N$ (so $v \neq x$). Since the $v_i y_i$ are all timelike, and $y_i \to x$, vx must be timelike or null. For, by 2.14, $\Phi(v_i, y_i) \to \Phi(v, x)$; so $\Phi(v, x) \geqq 0$ follows from $\Phi(v_i, y_i) \geqq 0$.

But vx cannot be timelike since $v \in \bar{F}$, $I^+[\bar{F}] = F$ (cf. 3.4) and $x \not\in F$. Thus vx is a null geodesic η. Furthermore, no point of η can lie in F (since $w \in F$ and $w \prec x$ would imply some $u \in F$ with $u \ll w \prec x$ so $u \ll x$ whence $x \in F$), whereas every point of each $v_i y_i$ lies in F. Hence $\eta \subset B$, as required.

3.20. THEOREM. *Let $S \subset M$ and set $B = \partial I^+[S]$. Then if $x \in B - \bar{S}$, there exists a null geodesic $\eta \subset B$ with future endpoint x and which is either past-endless or has a past endpoint on $\bar{S}$.*

Proof. Since $\bar{S}$ is closed and $x \notin \bar{S}$, we can choose an open set $Q \ni x$ not meeting $\bar{S}$. Condition (a) of 3.19 is clearly satisfied, so a null geodesic exists on B with future endpoint x. Define η to be the maximal extension of this geodesic into the past on B. Then if η is not past-endless it has a past endpoint y on B (since B is a closed set). If $y \notin \bar{S}$ we can apply 3.19 again to obtain another geodesic ζ on B with future endpoint y and which does not continue η. But by 2.19 this would lead to chronologically related points on B, contradicting the achronality of B.

3.21. *Remark.* Observe that the two possibilities for the null geodesic in 3.20 have been illustrated in our examples: in Fig. 19, a past endless null geodesic exists on B; whereas in Fig. 20, all null geodesics which are maximally extended on B, have past endpoints on $S = \eta\ (=\bar{S})$. Note also that in Fig. 20, the only place away from η at which two different null geodesics of B intersect, namely the point p, is a place at which a geodesic on B, if extended further, would have to leave the boundary B and enter the interior set $I^+[S]$. That this illustrates a general feature of achronal boundaries will be shown by the next proposition. There is also a version of the result for which the two intersecting null geodesics become "infinitesimally neighboring" geodesics on B. This result, which will have some importance to us later, will be given in Section 7 (cf. 7.27) after the concept of conjugate points has been discussed in detail.

3.22. PROPOSITION. *Let $B = \partial I^+[S]$. Suppose $x \in B - \bar{S}$ is an endpoint of two null geodesics on B. Then*:

(a) *if x is a past endpoint of one or both geodesics, then their union is a null geodesic on B*;

(b) *if x is a future endpoint of both geodesics, then unless one is contained in the other, every extension of either geodesic into the future beyond x must leave B and enter $I^+[S]$.*

Proof. To prove (a), suppose first that x is the past endpoint of one geodesic and the future endpoint of the other. Then, by 2.19 and the achronality of B, it follows that the union of the two null geodesics must be a single null geodesic as required. On the other hand, suppose x is the past endpoint of both null geodesics. By 3.20, another null geodesic having x as its future endpoint must exist on B. This must continue both geodesics, by the above remarks, so the union of all three is a single geodesic. To establish (b), suppose one of the geodesics to be extended, on B, into the future beyond x. By (a) the union of this extension ζ with the other geodesic must constitute a single null geodesic. This is impossible unless one of the two original geodesics contained the other. Excepting this situation, the geodesic extension ζ cannot lie on B, as required. On the other hand, since $\zeta \subset J^+(x)$, it follows from 3.9 that $\zeta - \{x\} \subset I^+[S]$.

SECTION 4

Global Causality Conditions

4.1. *Remark.* In 1.7 and 2.21, attention was drawn to the possibility that a space-time might possess closed trips ($x \ll x$) or closed causal trips ($x \prec y$, $y \prec x, x \neq y$). It is customary to dismiss such space-times, as models of the universe, on the grounds that they are unphysical, such gross causality violations leading to severe interpretive difficulties. The physical or philosophical reasons for ruling out such space-times are impressive. But perhaps they are not completely conclusive. In any case, it is often convenient to study space-times possessing causality violations, as part of a program of comprehending the global structure of space-time models in general. Thus, it is not necessary that all the models studied should necessarily be totally realistic in physical terms for them to have some indirect physical value. Also there are other types of causality violations possible, weaker than the existence of closed trips or closed causal trips. It is worthwhile to study some of these in conjunction with the ones just mentioned.

4.2. DEFINITION. A space-time M is *future-distinguishing at* $p \in M$ if and only if $I^+(p) \neq I^+(q)$ for each $q \in M$ with $q \neq p$; M is *future-distinguishing* if and only if it is future-distinguishing at every point. This property of being future-distinguishing is called *future-distinction.* The concept of *past-distinction* is defined similarly [18].

4.3. *Remark.* It is clear from 3.8 that no space-time containing closed causal trips can be either past- or future-distinguishing. However, in Fig. 21 a two-dimensional space-time is depicted which is future-distinguishing but not past-distinguishing and hence contains no closed causal trips.

4.4. DEFINITION. An open set $Q \subset M$ is *causally convex* if and only if Q intersects no trip in a disconnected set.[1] Let $p \in M$. Then M is *strongly causal at* p if and only if p has arbitrarily small causally convex neighborhoods. The space-time M is *strongly causal* if and only if it is strongly causal at every point [4].

4.5. *Remark.* "Arbitrarily small," in 4.4, means that such a neighborhood Q of p can be found inside any open set containing p (i.e., such Q's form a neighborhood base at p). Without the qualification "arbitrarily small," 4.4 would become vacuous since $Q = M$ is causally convex for any M. Observe, on the other hand, that for any M there are many arbitrarily small neighborhoods Q of p which are *not* causally convex. The "hour-glass" or even "spherelike" examples of Fig. 22 each illustrate this fact. However, such "local" violations of causal convexity

[1] Equivalently, the open set Q is causally convex if and only if for every $x, y \in Q$, $x \ll z \ll y$ implies $z \in Q$.

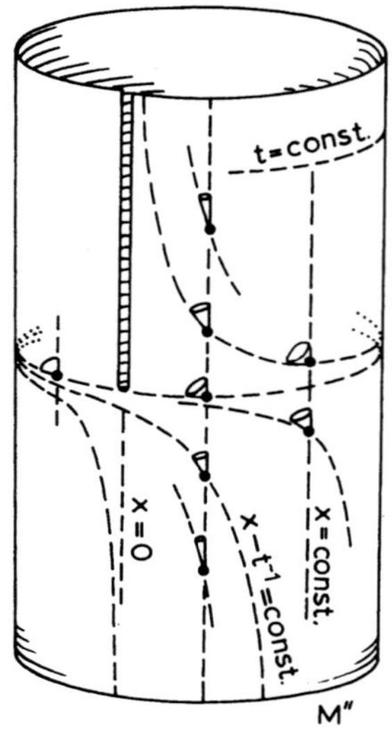

FIG. 21. *Consider the metric form* $ds^2 = dt\,dx + t^2\,dx^2$, *with* $\partial/\partial t$ *future-pointing, on the strip* $|x| \leqq 1$ *of the* (t, x)*-plane. Identifying* $(t, -1)$ *with* $(t, 1)$ *for each* t, *we obtain a space-time* M *with a closed causal trip (the null geodesic* $t = 0$*). Removal of the point* $(0, 0)$ *leaves us with a space-time* M' *with no closed causal trips, but which is neither future- nor past-distinguishing. (Take* p, q *on* $t = 0$, *then* $I^{\pm}(p) = I^{\pm}(q)$.*) If we remove the future-endless null geodesic* $t \geqq 0$, $x = 0$ *from* M, *we obtain a space-time* M'' *which is future-distinguishing but not past-distinguishing* [18]

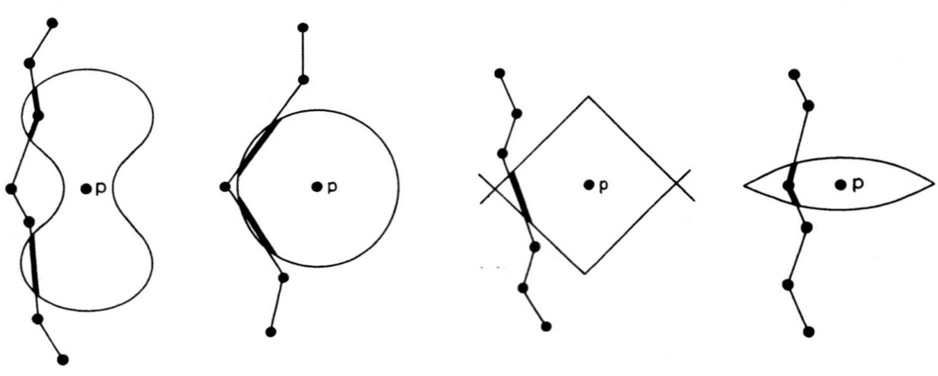

FIG. 22. *The two neighborhoods on the left are not causally convex; trips which intersect each in a disconnected set are depicted. Assuming no global connections not shown, the two neighborhoods on the right are causally convex*

are easily avoided, as the final two examples of neighborhoods depicted in Fig. 22 show. (This is made more explicit in 4.8.) Thus, a space-time M which violates strong causality at a point p must do so by virtue of its *global* structure. Roughly speaking, strong causality violation at p means that trips can leave the vicinity of p and then return to it, even though an actual closed trip or closed causal trip need not be the result. We shall see in 4.18 that a space-time which is strongly causal at p must be both future- and past-distinguishing at p. On the other hand, in Fig. 23, an example is given in which strong causality is violated even though the space-time is both future- and past-distinguishing.

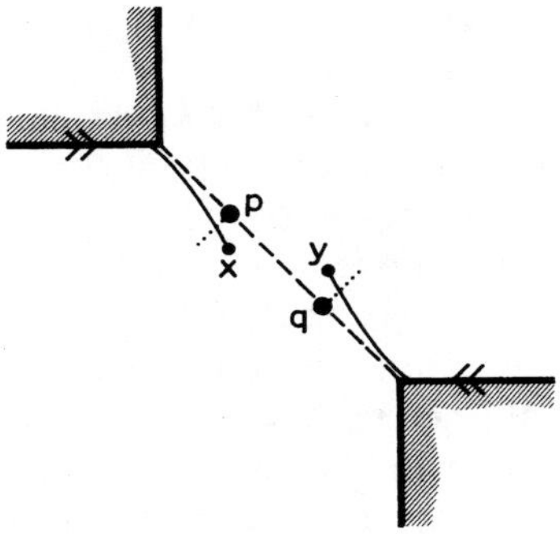

FIG. 23. *This identified subset of Minkowski 2-space is future- and past-distinguishing but strong causality fails all along the endless null geodesic through p and q. The Alexandrov topology is T_1 but not Hausdorff*

4.6. DEFINITION. Let Q be an open subset of M and let $x, y \in Q$. Then we write $x \ll_Q y$ if and only if a trip lying in Q exists from x to y, and $x \prec_Q y$ if and only if a causal trip in Q exists from x to y. Since Q is open it is a space-time manifold in its own right—or, if Q is not connected (and assuming $Q \neq \emptyset$), it is a disjoint union of space-time manifolds. Hence all the properties in 2.5 and 2.18 hold equally well for $\ll_Q$ and $\prec_Q$ as they do for $\ll$ and $\prec$.

Define:

$$\langle x, y \rangle_Q = \{z | x \ll_Q z \ll_Q y\}$$

and write

$$\langle x, y \rangle = \langle x, y \rangle_M$$

so that $\langle x, y \rangle = I^+(x) \cap I^-(y)$.

4.7. PROPOSITION. *The sets $\langle x, y \rangle$ are open; so are the sets $\langle x, y \rangle_Q$ if Q is open (with $x, y \in Q$).*

Proof. By 2.8, $I^+(x) \cap I^-(y) = \langle x, y \rangle$ is open in M. Thus, correspondingly, $\langle x, y \rangle_Q$ must be open in the space-time Q (or union of space-times Q). But Q is open in M, so $\langle x, y \rangle_Q$ is also open in M.

4.8. PROPOSITION. *If N is a simple region and $x, y \in N$, then the set $\langle x, y \rangle_N$ has the property that no trip (or causal trip) lying in N can intersect $\langle x, y \rangle_N$ in a disconnected set.*

Proof. Assume xy is future-timelike (otherwise $\langle x, y \rangle_N = \emptyset$, by 2.16). Let $\eta \subset N$ be a trip containing points $u, v \in \langle x, y \rangle_N$ with u preceding v along η. The portion of η from u to v, together with the timelike geodesics xu and vy (cf. 2.16) constitutes a trip from x to y which must (by the definition of $\langle x, y \rangle_N$) be contained in $\langle x, y \rangle_N$. This applies to any $u, v \in \eta \cap \langle x, y \rangle_N$, so $\eta \cap \langle x, y \rangle_N$ must be connected. The modification required if η is a causal trip is left as an exercise.

4.9. Proposition. *If N is a simple region, Q an open set contained in N and $p \in Q$, then there exist $u, v \in Q$ such that $p \in \langle u, v \rangle_N \subset Q$.*

Proof. Choose Minkowski normal coordinates for N with origin at p. Choose $\varepsilon > 0$ small enough that the whole of the normal coordinate ball B, given by $t^2 + x^2 + y^2 + z^2 \leqq \varepsilon^2$, is contained in Q and small enough that any timelike curve in B is also "timelike" with respect to the "flattened" Minkowski metric $4dt^2 - dx^2 - dy^2 - dz^2$. Take u at $(-\frac{1}{2}\varepsilon, 0, 0, 0)$ and v at $(\frac{1}{2}\varepsilon, 0, 0, 0)$. Then any timelike geodesic γ from u extends into the future in N until it meets the "hemisphere" defined by $t^2 + x^2 + y^2 + z^2 = \varepsilon^2$, $t > 0$. It must therefore cross the light cone of null geodesics with future endpoint v, since these describe a continuous hypersurface and extend into the past in B until they meet the opposite hemisphere defined by $t^2 + x^2 + y^2 + z^2 = \varepsilon^2$, $t \leqq 0$. If q is the intersection point of γ with this cone, then qv is future-null so no point r to the future of q on γ (or its extension in N) can have rv future-timelike (since $qr \cup rv$ would otherwise constitute a trip so, by 2.16, qv would have to have been future-timelike). Now suppose $w \in \langle u, v \rangle_N$. Then the geodesics uw and wv are future-timelike. Denoting uw (or its extension) by γ, we see by the above argument that w cannot lie to the future of q on γ, whence $w \in B$. Thus $\langle u, v \rangle_N \subset B \subset Q$.

4.10. Proposition. *Any simple region, if regarded as a space-time manifold in its own right, must be strongly causal.*

Proof. This follows at once from 4.8, 4.9 and the definition in 4.4.

4.11. Definition [4], [6]. A *local causality neighborhood* is a causally convex open set whose closure is contained in a simple region in M.

4.12. Proposition. *M is strongly causal at p if and only if p is contained in some local causality neighborhood.*

Proof. If M is strongly causal at p, choose a simple region $N \ni p$ and an open set $Q \ni p$ whose closure lies in N. A causally convex open set containing p exists in Q and is a local causality neighborhood as required. Conversely, suppose p belongs to a local causality neighborhood L contained in some simple region N. By 4.9, we can find arbitrarily small sets $\langle u, v \rangle_N \subset L$ containing p. If a trip γ in M were to intersect $\langle u, v \rangle_N$ in a disconnected set, then by 4.8, $\gamma \not\subset N$. In fact, γ would clearly have to leave and re-enter N, indeed, to leave and re-enter L. But this would contradict the causal convexity of L. Hence $\langle u, v \rangle_N$ is causally convex, so M is strongly causal.

4.13. Proposition. *The set of points at which M is strongly causal is open.*

Proof. Immediate from 4.12.

4.14. Proposition. *Let $A \subset M$ and suppose that strong causality holds at every point of A. Then A can be covered by a locally finite (countable) system of local causality neighborhoods. If A is compact, then a finite number of such neighborhoods will suffice.*

Proof. This follows from 4.12 and the paracompactness of M [35].

4.15. Proposition. *No local causality neighborhood can contain a future- or past-endless causal trip.*

Proof. Suppose a local causality neighborhood L contains a future-endless causal trip γ. Let $p_1, p_2, p_3, \cdots$ be a sequence of points proceeding indefinitely along γ (i.e., if $q \in \gamma$, then some p_i lies to the future of q on γ). Then since L has compact closure, $\bar{L}$ being contained in a simple region N (cf. 1.12), there must be an accumulation point $p \in \bar{L}$ of the $\{p_i\}$. We have $\bar{L} \subset N$, so $p \in N$. Now p is not a future endpoint of γ. Thus, there exists a neighborhood Q of p such that there are points arbitrarily far into the future along γ not contained in Q (cf. 1.8). Choose $u, v \in Q$ so that $p \in \langle u, v\rangle_N \subset Q$ (cf. 4.9). Then $\langle u, v\rangle_N$ contains infinitely many point p_i on γ but also fails to contain infinitely many points on γ between p_i's. This contradicts 4.8.

4.16. Lemma [18]. *Let $p \in M$. Then strong causality fails at p if and only if there exists $q \prec p$, with $q \neq p$, such that: $x \ll p$ and $q \ll y$ together imply $x \ll y$ for all x, y.*

Proof. Suppose strong causality fails at p. Let N be a simple region containing p and let $Q_i = \langle u_i, v_i\rangle_N$ be a nested sequence of neighborhoods of p converging on p: $Q_1 \supset Q_2 \supset Q_3 \supset \cdots, \bigcap_i Q_i = \{p\}$. We may take $\bar{Q}_i \subset N$; then each Q_i must fail to be causally convex since it would otherwise be a local causality neighborhood, violating 4.12. Let γ_i intersect Q_i in a disconnected set. By 4.8, $\gamma_i \not\subset N$. We can take γ_i to have a past endpoint a_i in Q_i and to exit N first at $b_i \in \partial N$, finally to re-enter N at $c_i \in \partial N$ and to terminate with future endpoint $d_i \in Q_i$ (see Fig. 24).

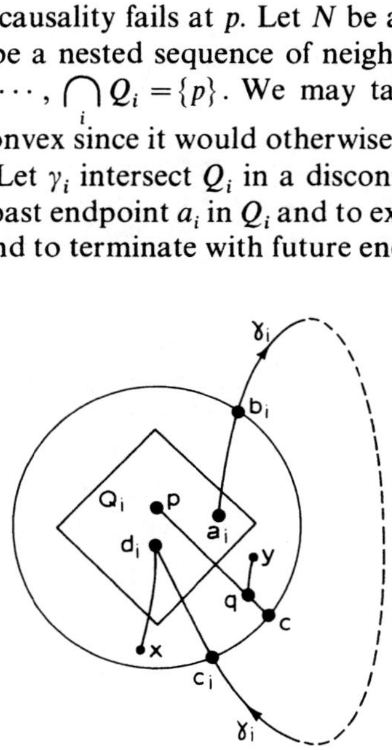

Fig. 24. *Diagram for the proof of* 4.16

(Possibly γ_i could have other portions in common with N as well.) Let $c \in \partial N$ be an accumulation point of $\{c_i\}$ (∂N is compact). The geodesics $c_i d_i$ are future-timelike. Hence (cf. 2.14) cp must be future-causal. Choose q between c and p on cp. Now suppose $x \ll p$ and $q \ll y$. Since $p \in I^+(x)$ and $I^+(x)$ is open, it follows that $Q_i \subset I^+(x)$ for large enough i, so that $a_i \in I^+(x)$.

Furthermore, $c \prec q \ll y$ implies $c \in I^-(y)$ (cf. 2.18), whereas $I^-(y)$ is open so $c_i \in I^-(y)$ for infinitely many values of i. Then we have, for some large enough i, $x \ll a_i \ll b_i \ll c_i \ll y$, so $x \ll y$ as required.

For the converse, assume $q \prec p$ and that $x \ll p$ and $q \ll y$ together imply $x \ll y$. Let $P \ni p$ and $Q \ni q$ be disjoint open sets. I shall show that P cannot be causally convex no matter how small it is chosen. Take $x \in P \cap I^-(p)$ and

$y \in Q \cap I^+(q) \cap I^-(z)$, where $z \in P \cap I^+(p)$. (Clearly $P \cap I^+(p) \neq \emptyset$. We have $q \prec p \ll z$ so $q \in I^-(z)$. But $I^-(z)$ is open, so $Q \cap I^+(q) \cap I^-(z) \neq \emptyset$ also.) Choose a trip from x to z via the point y. (We have $x \ll y$ and $y \ll z$.) This trip clearly meets P in a disconnected set.

4.17. *Remark.* The time-reverse of 4.16 also holds, so we have another condition equivalent to strong causality failure at p (strong causality being a time-symmetric condition). Note that 4.16 (and its time-reverse) imply that strong causality cannot just fail at only a single point of M; for if strong causality fails at p, then it must fail at q also. (This point is elaborated in 4.31.) The condition in 4.16 can also be rephrased in numerous other ways. For example, strong causality fails at p if and only if there exists a point $q \in J^-(p) - \{p\}$ such that $I^+(x) \supset I^+(q)$ for all $x \in I^-(p)$. (It is worth examining these various aspects of 4.16 in relation to Fig. 23.) Note that if $q \ll p$ in 4.16, then there are closed trips through p (exercise).

4.18. PROPOSITION. *If M is strongly causal at p, then M is future-distinguishing at p.*

Proof. Suppose $I^+(p) = I^+(q)$ for some $q \neq p$. As in the final argument in the proof of 4.16, let $P \ni p$ and $Q \ni q$ be disjoint and open. Choose $x \in I^+(p) \cap P$. Then $q \ll x$. Choose y in Q with $q \ll y \ll x$. Then $p \ll y$. Thus there is a trip from p to x via $y \notin P$, which intersects P in a disconnected set. This holds for arbitrarily small P, so strong causality must fail at p. (The result can also be proved rather rapidly using 3.19 and 4.16. This is left as an exercise for the reader.)

4.19. *Remark.* We have seen that various degrees of causality restriction on a space-time are possible (e.g., in order of decreasing restrictiveness: strong causality, future- and past-distinction, future-distinction, absence of closed causal trips, absence of closed trips). Each of these restrictions may be regarded as "reasonable" from the physical point of view since if any one of the conditions is violated for a space-time M, it is possible to modify the metric of M, in some compact region, by an arbitrarily small amount, so as to produce a space-time with closed trips. However, there are many other causality restrictions also with this property. A number of inequivalent conditions, each more restrictive than strong causality, have been suggested by Carter [24]. For example, given an integer $n \geqq 2$, we may require that for any selection of n distinct points $p_1, p_2, \cdots, p_n \in M$ there should exist arbitrarily small neighborhoods $Q_i \ni p_i$, $i = 1, \cdots, n$, such that it is impossible to find n trips $\gamma_1, \gamma_2, \cdots, \gamma_n$ for which the past endpoint of γ_i lies in Q_i and the future endpoint of γ_i lies in Q_{i+1} if $i \neq n$ and in Q_1 if $i = n$. Examples (due to Carter) can be constructed which violate this condition for one value of n but satisfy it for all smaller values of n. One such can be obtained by taking the n-fold covering space of the space-time of Fig. 23. (For another example see Fig. 25.)

In the face of this, it is fortunate that a "maximally restrictive" causality condition exists which is acceptable on physical grounds. This is Hawking's notion of *stable causality* [25]. A space-time is stably causal if it cannot be made to contain closed trips by arbitrarily small perturbations of the metric. The precise formulation of this is best carried out in terms of the bundle of metrics over a manifold, but I shall not enter into this here. I merely remark that Hawking has shown that the condition of stable causality is equivalent to the existence of a

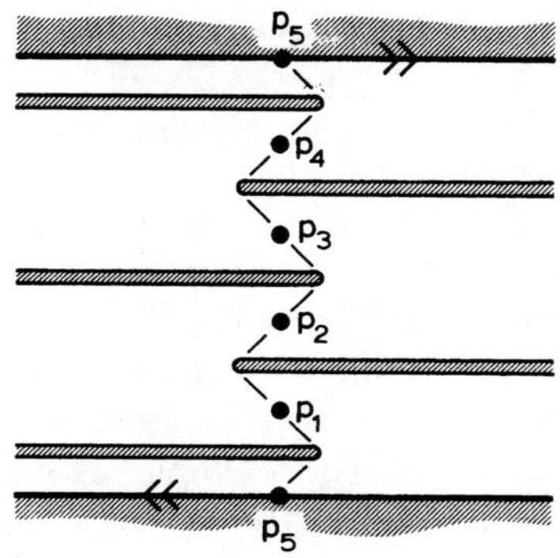

FIG. 25. *Delete the lines* $t = 2k - 1, (-1)^k x + 1 \geqq 0, k = 1, 2, \cdots, n$, *and the regions* $t \leqq 0$ *and* $t \geqq 2n$ *from Minkowski 2-space. Identify* $(0, x)$ *with* $(2n, (-1)^n x)$. *The resulting space-time violates Carter's causality condition for the value n, but satisfies it for* $(n - 1)$ (*if* $n \geqq 2$)

global time function on M, that is to say, a scalar field t on M whose gradient $\nabla^a t$ (i.e., $g^{-1}\,dt$) is everywhere timelike and future-pointing [25].

4.20. PROPOSITION. *If* $p \ll q$ *and* $p \ll r$, *then there exists a point* w *such that* $w \ll q$, $w \ll r$ *and* $p \ll w$.

Proof. If $p \ll q$ and $p \ll r$, then $p \in I^-(q) \cap I^-(r)$, which is open. Hence there is a point w in $I^-(q) \cap I^-(r)$ lying just to the future of p on some trip from p.

4.21. PROPOSITION. *If* $x, p, q, r, s \in M$ *are such that* $x \in \langle p, q \rangle \cap \langle r, s \rangle$, *then there exist* $u, v \in M$ *such that* $x \in \langle u, v \rangle \subset \langle p, q \rangle \cap \langle r, s \rangle$.

Proof. We have $x \ll q$, $x \ll s$, so by 4.20 there is a point v with $x \ll v$, $v \ll q$, $v \ll s$. Similarly, using the time-reverse of 4.20 we obtain a point u with $u \ll x$, $p \ll u$, $r \ll u$.

4.22 DEFINITION. The property of 4.21, together with the fact that any $p \in M$ is contained in some set $\langle u, v \rangle$, shows that the sets of the form $\langle u, v \rangle$ constitute a base for a topology on M. This is called the *Alexandrov topology*. (That is to say, an open set in the Alexandrov topology is a union of sets of the form $\langle u, v \rangle$.)

4.23. *Remark.* This topology is sometimes called the *interval topology* of the space-time (cf. Pimenov [26]). The terminology I am adopting here follows that of Kronheimer and Penrose [18]. A. D. Alexandrov appears to have been the first to suggest basing the topological properties of a space-time solely on its causality structure [27]. The necessary property of 4.20, together with its time-reverse, and the fact that each point has a chronological successor and predecessor, is called *fullness* in the context of causal space theory (cf. 2.21) [18].

Clearly the Alexandrov topology agrees with the manifold topology in the case of Minkowski space. It is also clear that the two topologies must differ for any space-time with closed trips. (For example, if M is the portion $0 \leqq t \leqq 1$ of Minkowski space, where $(0, x, y, z)$ is identified with $(1, x, y, z)$, then every set $\langle u, v \rangle$ is the whole of M.) The next theorem gives the complete condition that the two topologies should agree [18].

4.24. THEOREM. *The following three restrictions on a space-time M are equivalent:*
(a) *M is strongly causal;*
(b) *the Alexandrov topology agrees with the manifold topology;*
(c) *the Alexandrov topology is Hausdorff.*

Proof. First, (a) implies (b). To show this, we need only establish that by virtue of (a), every open set in the manifold topology is open in the Alexandrov topology. (The fact that Alexandrov open sets are open in the manifold topology is obvious by 4.7.) Now suppose strong causality holds at p and P is an open set (in the manifold topology) containing p. We have to show that an Alexandrov neighborhood containing p exists in P. Let N be a simple region in P containing p and let $Q \ni p$ be a causally convex open set contained in N (which exists because of the strong causality). By 4.9 we have $u, v \in Q$ such that $p \in \langle u, v\rangle_N \subset Q$. But if $\langle u, v\rangle_N \neq \langle u, v\rangle$ this can only be because of the existence of a trip from u to v which leaves and re-enters N. Thus it would have to leave and re-enter Q also, violating the causal convexity of Q. Thus, $p \in \langle u, v\rangle \subset Q \subset P$ as required.

The fact that (b) implies (c) is obvious, since M was assumed to be Hausdorff in its manifold topology. It remains to show that (c) implies (a). Suppose that (a) is false and strong causality is violated at p. Let $q \prec p$ be as in 4.16. I shall show that any Alexandrov neighborhood of p must intersect every Alexandrov neighborhood of q, so that the Alexandrov topology fails to be Hausdorff, as required. Let $p \in \langle x, u\rangle$ and $q \in \langle v, w\rangle$. We have $q \prec p \ll u$, so $q \in I^-(u)$. Choose y just to the future of q, giving $q \ll y$, $y \in I^-(u)$ and $y \in \langle v, w\rangle$. By 4.16 we have $x \ll y$, so $y \in \langle x, u\rangle$ also. Thus $\langle x, u\rangle \cap \langle v, w\rangle \neq \varnothing$.

4.25. *Remark.* It is worthwhile to examine Fig. 23 again to see how the failure of the Hausdorff condition for the Alexandrov topology arises here. In fact the example in Fig. 23 illustrates the fact that it is the Hausdorff condition (i.e., distinct points have disjoint neighborhoods) rather than, say, the weaker T_1 condition (i.e., for every pair of distinct points, a neighborhood of each exists which does not contain the other) which is relevant. Actually, the Alexandrov topology of Fig. 23 is T_1 (but not Hausdorff), as is not hard to verify. An example whose Alexandrov topology is *not* T_1 (and therefore also not Hausdorff) but for which the space-time is still both future- and past-distinguishing is illustrated in Fig. 26.

Notice that in each of Figs. 21, 23, 26 there is a null geodesic along which strong causality fails. This is actually one aspect of a general result concerning the region of strong causality failure in a space-time (cf. 4.31). I shall devote the remainder of this section to certain properties relating to the structure of this region.

4.26. DEFINITION. A point $p \in M$, through which passes a closed trip, is called *vicious* (Carter [24]). Denote the set of all vicious points of M by the letter V. We clearly have

$$V = \bigcup_{x \in M} \langle x, x\rangle.$$

Each $\langle x, x\rangle$ is open, by 4.7, so V is open.

4.27. PROPOSITION [24]. *If $\langle x, x\rangle \cap \langle y, y\rangle \neq \varnothing$, then $\langle x, x\rangle = \langle y, y\rangle$. Hence, V is a union of disjoint open sets of the form $\langle x, x\rangle$.*

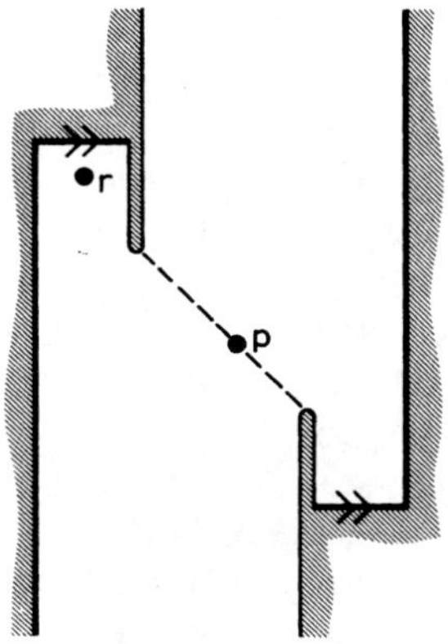

FIG. 26. *This example is both future- and past-distinguishing but its Alexandrov topology is not T_1. For, any Alexandrov neighborhood of p contains r. The Alexandrov topology is therefore not Hausdorff either, the space-time being not strongly causal*

Proof. Suppose $z \in \langle x, x\rangle \cap \langle y, y\rangle$. Then if $w \in \langle y, y\rangle$, we have $x \ll z \ll y \ll w \ll y \ll z \ll x$, so $w \in \langle x, x\rangle$. Thus $\langle y, y\rangle \subset \langle x, x\rangle$. Similarly $\langle x, x\rangle \subset \langle y, y\rangle$, so $\langle x, x\rangle = \langle y, y\rangle$.

4.28. PROPOSITION. $\partial V = \bigcup_{x \in V} \partial\langle x, x\rangle$.

Proof. This follows since the $\langle x, x\rangle$'s are open and disjoint.

4.29. PROPOSITION. *Strong causality fails at each point of ∂V.*

Proof. Obvious from the definition in 4.4.

4.30. PROPOSITION. *If future-distinction fails at $p \notin V$, then p lies on a past-endless null geodesic $\gamma \subset \sim V$ along which future-distinction fails (so $I^+(q) = I^+[\gamma]$ for each $q \in \gamma$).*

Proof. Suppose M is not future-distinguishing at p ($\notin V$) and consider $\partial I^+(p)$. Since $p \notin I^+(p)$ ($p \notin V$) and $I^+(p) \subset I^+(p)$ we have, by 3.4, $p \in \partial I^+(p)$. Furthermore, $I^+(p) = I^+(q)$ for some $q \neq p$. We can apply 3.19 to $B = \partial I^+(p) = \partial I^+(q)$ to obtain a null geodesic γ on $\partial I^+(p)$ which extends indefinitely into the past from its future endpoint p. (We can clearly choose the Q of 3.19 to avoid one or other of p, q.) Now if $r \in \gamma$ we have $r \prec p$, so $I^+(r) \supset I^+(p)$; also we have $r \in \partial I^+(p)$ so $I^+(r) \subset I^+(p)$ (and $r \notin I^+(p)$). Thus $I^+(r) = I^+(p)$, so future-distinction fails at each point of γ. Furthermore $r \notin I^+(p) = I^+(r)$, so $r \notin V$, whence $\gamma \subset \sim V$.

4.31. THEOREM. *Suppose strong causality fails at p. Then at least one of the following holds:*

(a) *there are closed trips through p (i.e., $p \in V$);*

(b) *p lies on a past-endless null geodesic on ∂V, at every point of which future-distinction fails;*

(c) *p lies on a future-endless null geodesic on ∂V, at every point of which past-distinction fails;*

(d) *p lies on both a past-endless null geodesic on ∂V along which future-distinction fails and a future-endless null geodesic on ∂V along which past-distinction fails, except that at p itself M may be both past- and future-distinguishing;*

(e) *an endless null geodesic* γ *through* p *exists, at every point of which strong causality fails, such that if* u *and* v *are any two points of* γ *with* $u \prec v$, $u \neq v$, *then* $u \ll x$ *and* $y \ll v$ *together imply* $y \ll x$.

Proof. Let N be a simple region containing p and let $Q_i = \langle u_i, v_i \rangle_N$ be a nested sequence of neighborhoods of p converging on p. Then (as in the proof of the lemma in 4.16, cf. Fig. 24) for each i $(=1, 2, 3, \cdots)$ there exists a trip γ_i from a point $a_i \in Q_i$, which first exits N at a point $b_i \in \partial N$, which re-enters N for the last time at a point $c_i \in \partial N$, and which terminates with future endpoint $d_i \in Q_i$. Let the pair (b, c) be an accumulation point of (b_i, c_i) on $\partial N \times \partial N$ (which is compact). Then pb and cp must be future-causal (since pb_i and c_ip are all future-timelike). Now, various possibilities can occur. Suppose first that each of pb and cp is timelike. Then, for some i, $b_i \in I^+(p)$ and $c_i \in I^-(p)$, so $p \ll b_i \ll c_i \ll p$. This is case (a). Secondly, suppose instead that pb is timelike but cp is null. Let $x \in \langle p, b \rangle_N$. Then $c \ll x$, so for some large enough i we have $c_i \ll x$ together with $x \ll b_i$. But $b_i \ll c_i$, so $x \in V$. This yields $\langle p, b \rangle_N \subset V$, whence $p \in \bar{V}$. Assume $p \notin V$, since the other possibility has been already considered. Then $p \in \partial V$. Now any $y \in \langle c, x \rangle$ satisfies $p \ll x \ll b_i \ll c_i \ll y$ for some large enough i, whence $I^+(c) \subset I^+(p)$. But $c \prec p$ implies $I^+(c) \supset I^+(p)$, so $I^+(c) = I^+(p)$ and past-distinction fails at p. Furthermore, the γ of 4.30 may be chosen to be the maximal extension of pc into the past. We have $\gamma \subset \sim V$, by 4.30. But $q \in \gamma$ implies $I^+(q) = I^+(p)$, so any point $z \in \langle q, x \rangle$ must lie on a trip from p, giving $x' \ll z$ for some $x' \in \langle p, b \rangle_N$. It is clear from 4.27 that every point of $\langle p, b \rangle_N$ lies in the same set $\langle x, x \rangle$, so $z \ll x \ll x' \ll z$, giving $z \in V$. Thus $\langle q, x \rangle \subset V$, whence $q \in \partial V$ as required for (b).

Thirdly we can suppose that pb is null but cp is timelike. This is the time-reverse of the previous case, so we obtain (c) (or (a)). Fourthly, suppose that both cp and pb are null, but that their directions differ, so that cb is timelike. Any point $x \in \langle c, b \rangle_N$ satisfies $x \ll b_i \ll c_i \ll x$, for some i, showing that $\langle c, b \rangle_N \subset V$. Thus $p \in \langle c, b \rangle_N \subset \bar{V}$. We may suppose $p \notin V$, otherwise we have (a) again. Thus $p \in \partial V$. Choose any point r on pb (with $p \neq r \neq b$). Consider the possibility $r \in V$. In this case some closed trip from r to r exits from N first at w, say. Then rw is future-timelike and so is pw. It is clear from 4.27 that every point of $\langle c, b \rangle_N$ belongs to the same set $\langle x, x \rangle$, so there are trips from w to points arbitrarily close to b. It follows that we are in the situation leading to case (b) above, where w takes over the role of b. Similarly, if a point of cp lies in V, then we have case (c). So we may suppose that cp and pb each lie on ∂V. Then any point p' between c and p on cp, if used in place of p, will satisfy the conditions leading to case (b) above, $p'b$ being future-timelike. Similarly if p'' lies between p and b we get case (c) above with p'' in place of p. It follows that we are in situation (d) for the point p.

Finally, suppose that cp and pb are both null, being portions of a single endless null geodesic γ. Extend each of c_id_i maximally as a timelike geodesic η_i, where the portion from c_i in $\bar{N}$ terminates at $e_i \in \partial N$. Similarly extend a_ib_i maximally as a timelike geodesic ζ_i, with portion f_ib_i in $\bar{N}$, where $f_i \in \partial N$. We have $f_i \ll b_i \ll c_i \ll e_i$ showing that strong causality fails at b and c and at every point between b and c on bc also. We can repeat the construction with b in place of p and p in place of c to obtain a new point b' in place of b. If $b' \notin \gamma$ we have $p \ll b'$ and we could have chosen some point $b'' \in \partial N$ on the trip from p to b' in our original construction, in place of b. (We would have $b' \in I^+(b'')$ so $I^+(b'')$ would contain infinitely many of the b_i''.) This would give us case (b) again. Thus we may take $b' \in \gamma$. Repeating the construction indefinitely in the future and past we see that strong causality failure may be assumed at every point of γ and examining the construction, we see that for large enough i the timelike geodesics η_i and ζ_i can be used to supply the required points in the neighborhoods of points of γ. That is to say, if $u, v \in \gamma$ (taking $u \prec v$ and $u \neq v$) and if U and W are neighborhoods of u and v, respectively, then η_i has a point m_i in U and ζ_i has a point n_i in W, such that $n_i \ll m_i$. If $u \ll x$ and $y \ll v$, choose $U \subset I^-(x)$ and $W \subset I^+(y)$. Then $y \ll n_i \ll m_i \ll x$.

4.32. *Remark.* The theorem in 4.31 yields only a part of the information which can be inferred concerning the structure of the region of strong causality violation (cf. Carter [24]). But let me leave it at that, save to illustrate the situation with a number of examples. In Fig. 27 each of the situations (a), (b), (c), (d) is illustrated

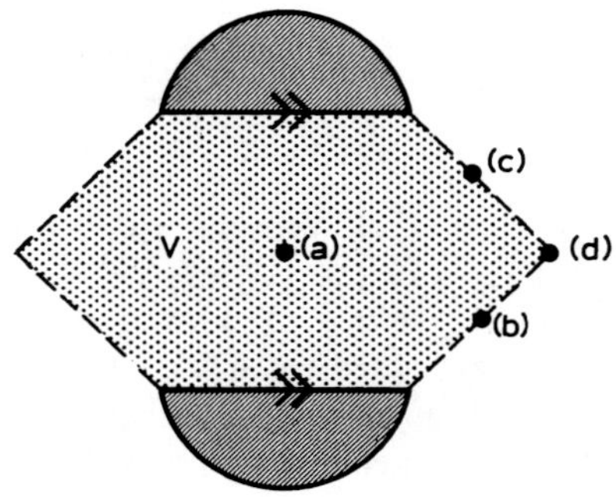

FIG. 27. *For points situated as indicated, different parts of* 4.31 *are illustrated*

in a simple example. (Case (e) has already been illustrated in Figs. 21, 23, 26.) In Fig. 28, p satisfies (d) twice over, but not (e) since the last part is not satisfied.

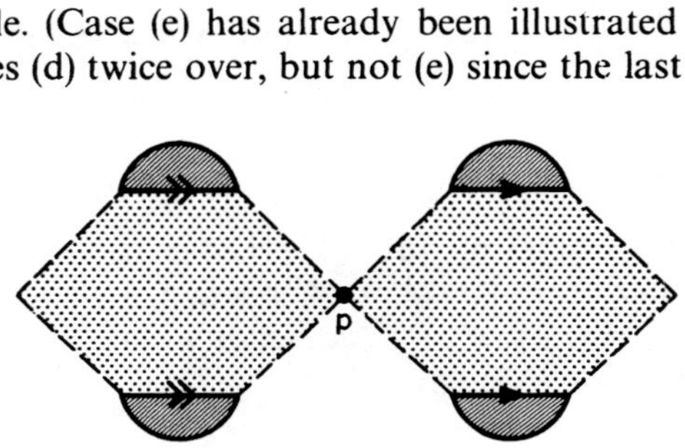

FIG. 28. *Here p illustrates part* (d) *in* 4.31 *twice over, but not* (e), *despite the existence of two endless null geodesics of strong causality failure through p*

Figure 29 is *apparently* similar but now both (d) and (e) hold (each twice over). Here, V consists of a single $\langle x, x \rangle$ whereas in the previous example V was the union of two such sets. In Fig. 30 p satisfies (b), (c) and (d), but not (e). An example which shows that not every point of ∂V need satisfy (b), (c), or (d) is given by the

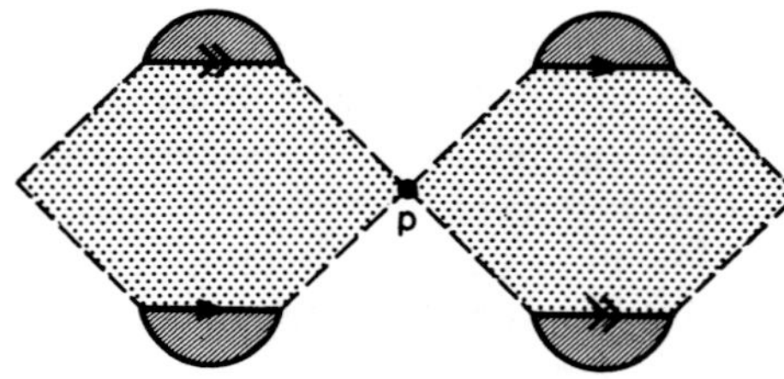

FIG. 29. *Apparently similar to Fig.* 28, *but now* (e) *holds at p as well as* (d) (*each twice over*)

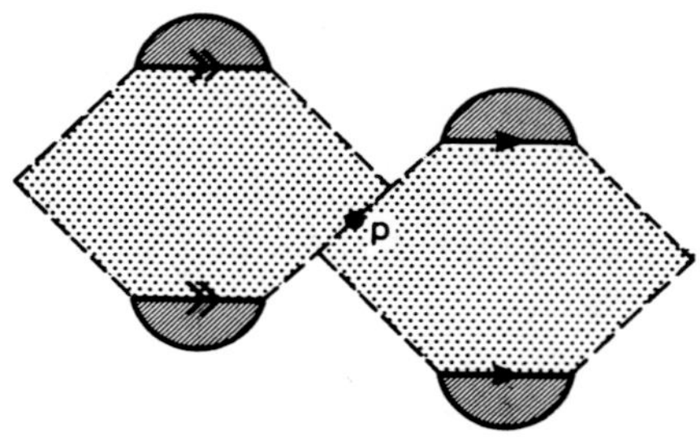

FIG. 30. *Here p illustrates all of* (b), (c), (d) *in* 4.31, *but not* (e)

following. Let M be the portion of Minkowski space given by $|t| \leqq 1$ where we delete $(x + a)^2 + y^2 + z^2 \leqq (1 + a)^2$, $t = 1$ and $(x - a)^2 + y^2 + z^2 \leqq (1 - a)^2$, $t = -1$ and where we identify $(1, x, y, z)$ with $(-1, -x, y, z)$ whenever these inequalities on x, y and z are not satisfied. If $a > 0$, then the portion of the null geodesic $t - x = y = z = 0$ for which $|t| \leqq a$, consists of points of ∂V for which (e) holds but not (c) or (d). On this geodesic we have (b) holding if $t < -a$ and (c) holding if $t > a$.

The following example, due to Carter [24], shows that the region of strong causality violation can be *compact* even though there are no closed causal trips. Here M is described by coordinates (t, y, z) of unrestricted range but with (t, y, z) identified with $(t, y + m, z + \pi n)$ for each pair of integers (m, n). This gives M the topology $\mathbb{R}^1 \times S^1 \times S^1$. The metric is taken to be

$$ds^2 = (\cosh t - 1)^2(dt^2 - dy^2) + dt\,dy - dz^2.$$

Strong causality failure occurs on the torus $t = 0$. In fact M is neither future- nor past-distinguishing on $t = 0$, although this fact is by no means self-evident. It is not possible for a space-time to be compact without possessing closed trips, as the following proposition shows (cf. also [26], [28], [29]).

4.33. PROPOSITION. *If M is compact it contains closed trips.*

Proof [18], [5]. Since every Alexandrov neighborhood $\langle x, y \rangle$ is open in the manifold topology, it suffices to show that compactness in the Alexandrov topology implies the existence of closed trips. Assume that M can be covered by a finite number of sets $\langle x_i, y_i \rangle$. Then for each y_i there is a j such that $y_i \in \langle x_j, y_j \rangle$, so $y_i \ll y_j$. Thus we have an infinite succession: $y_{i_1} \ll y_{i_2} \ll y_{i_3} \ll \cdots$. Since there are only a finite number of y_i's, there must be repetitions in the list and therefore closed trips in M.

SECTION 5

Domains of Dependence

5.1. DEFINITION. Let S be an achronal subset of M. Define the future and past domains of dependence of S and the total domain of dependence of S, respectively, as follows:

$$D^+(S) = \{x | \text{every past-endless trip containing } x \text{ meets } S\},$$

$$D^-(S) = \{x | \text{every future-endless trip containing } x \text{ meets } S\},$$

$$D(S) = \{x | \text{every endless trip containing } x \text{ meets } S\}.$$

Clearly $D(S) = D^+(S) \cup D^-(S)$.

5.2. *Remark.* A number of examples illustrating domains of dependence are given in Figs. 31–34. The significance of this notion from the point of view of physics is, roughly speaking, that $D(S)$ represents the region of space-time throughout which the physical situation would be expected to be determined, given suitable initial data on an achronal set S. This is assuming that the local physical laws are of a suitable "deterministic" and "causal" nature (being locally "Lorentz covariant," so that the bicharacteristics of the partial differential equations involved should be null geodesics in the space-time). One can envisage that physical

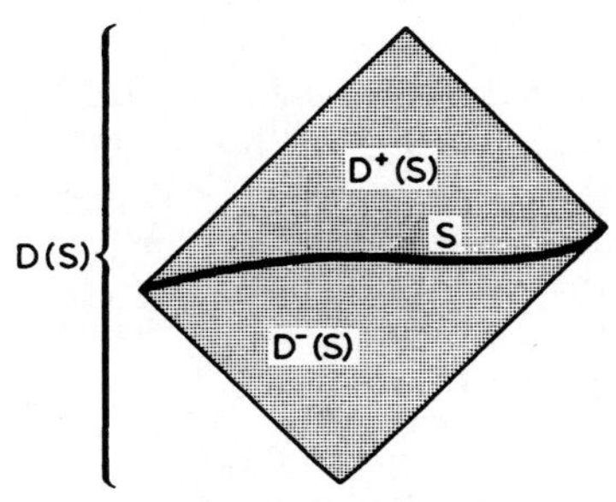

FIG. 31. *The domains of dependence of an achronal set S*

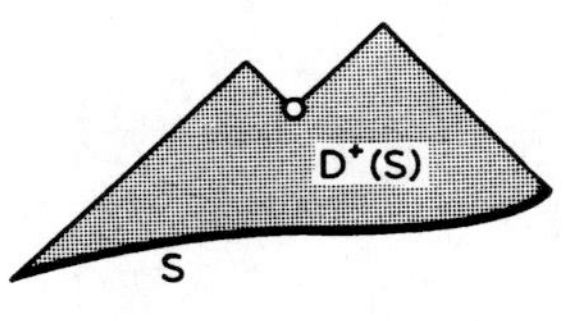

FIG. 32. *The effect on $D^+(S)$ of removing a point from the manifold M*

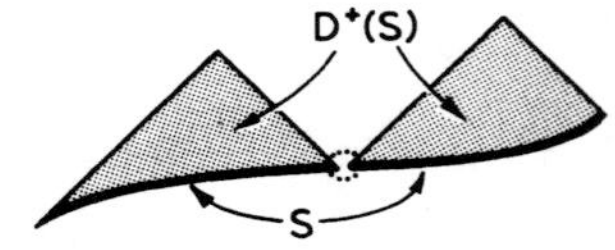

FIG. 33. *If one point is removed fron the achronal set S, the effect on $D^+(S)$ is similar to that which would have been obtained by removing the point from the manifold M*

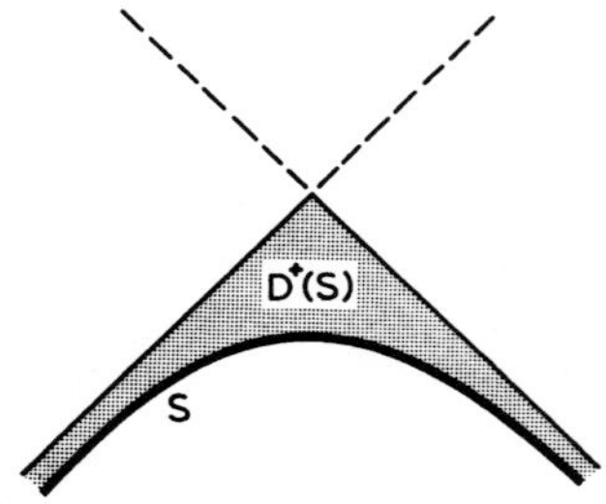

FIG. 34. *If S is the (geodesically complete) spacelike hypersurface* $t = -(x^2 + y^2 + z^2 + 1)^{1/2}$ *in Minkowski space, then* $D^+(S)$ *is* $(x^2 + y^2 + z^2)^{1/2} \leqq -t \leqq (x^2 + t^2 + z^2 + 1)^{1/2}$

information can be carried along timelike curves. Thus if a past-endless timelike curve, not meeting S, can terminate at a point p $(\in I^+[S])$, then we can imagine that information can be carried in from infinity along γ to influence the physics at p, this information being not taken account of by the data on S. This is essentially the situation which is prevented from occurring if $p \in D^+(S)$. These statements are all somewhat vague. But the physical significance of $D^\pm(S)$ is not really what will concern us here. These sets are useful as mathematical constructs quite independently of their interpretation.

We may ask whether a definition of domain of dependence given in terms of timelike curves rather than trips would be equivalent to the one given in 5.1. For an achronal (or closed) S the definitions are in fact equivalent (proof: exercise), but not if we do not restrict S in some such way (cf. Fig. 35). It does not appear to be generally useful to define $D^+(S)$ when S is not achronal (and the physical

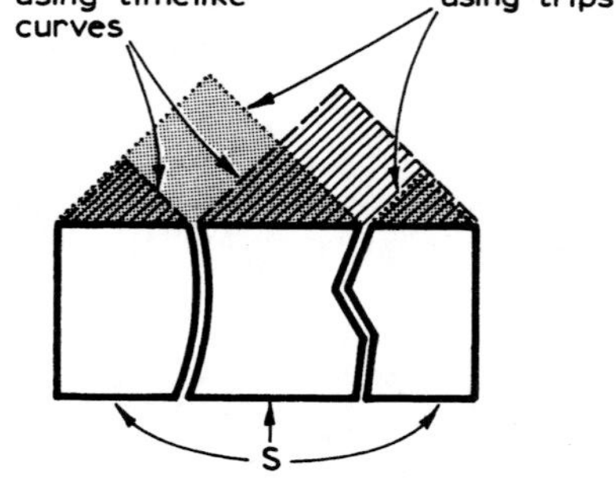

FIG. 35. *If S is neither achronal nor closed, it can make a difference whether trips or timelike curves are used in the definition of $D^+(S)$*

motivation largely disappears in such cases). For simplicity, one may also normally restrict attention to the case when S is closed. (One could argue that physical initial data should be continuous, so its value on an achronal set should define its value on the closure of the set, but this would rule out the use of Dirac δ-functions as initial data, so the issue is not clear.)

One could also use *causal* trips (or causal curves) to define domains of dependence (and there would be some "physical" justification for this). This would lead to certain minor differences from the theory described here. Some other authors have preferred this alternative choice (cf. Hawking [2]–[5]). My restriction of attention to trips rather than causal trips in 5.1 has the effect of keeping the theory relatively simple since $D^+(S)$ is then always closed provided S is closed (cf. 5.5(a)). As a general rule, properties based on trips are easier to handle than those based on causal trips.

I shall tend not to go into quite so much detail henceforth, as in the earlier sections. (In fact, a readable account by Geroch [30] of much of the material of this section already exists in the literature.) It is hoped that the reader will by now have gained some facility with the basic techniques, so less detail may be necessary than before.

5.3. DEFINITION. The future, past, or total Cauchy horizon of an achronal closed set S is defined as (respectively):

$$H^+(S) = \{x | x \in D^+(S) \text{ but } I^+(x) \cap D^+(S) = \emptyset\},$$

$$H^-(S) = \{x | x \in D^-(S) \text{ but } I^-(x) \cap D^-(S) = \emptyset\},$$

$$H(S) = H^+(S) \cup H^-(S).$$

The definitions of $H^\pm(S)$ can be restated as:

$$H^\pm(S) = D^\pm(S) - I^\mp[D^\pm(S)].$$

5.4. *Remark.* The future Cauchy horizon of S may be described as the future boundary of the future domain of dependence of S. In the example of Fig. 34, $H^+(S)$ is the set $t = -(x^2 + y^2 + z^2)^{1/2}$ and $H^-(S)$ is empty. If S is the hyperplane $t = 0$ in Minkowski space, then both of $H^\pm(S)$ are empty. In these cases $H(S)$ has no point in common with S. However it is often the case that $H(S)$ and S do have points in common. An example is given when S is the past light cone $t = -(x^2 + y^2 + z^2)^{1/2}$ in Minkowski space. Then $H^+(S) = S$ and $H^-(S) = \emptyset$. If S is the null hyperplane $t = x$, then $H^+(S) = S = H^-(S)$. If S is the ball $x^2 + y^2 + z^2 \leqq 1$, $t = 0$, then $H^\pm(S)$ is $(x^2 + y^2 + z^2)^{1/2} \pm t = 1, 0 \leqq \pm t$, its intersection with S being the sphere $x^2 + y^2 + z^2 = 1$, $t = 0$.

5.5. PROPOSITION. *Let $S \subset M$ be achronal and closed. Then*:

(a) *$D^+(S)$ is closed,*

(b) *$H^+(S)$ is achronal and closed,*

(c) $S \subset D^+(S)$,

(d) *$x \in D^+(S)$ implies $I^-(x) \cap J^+[S] \subset D^+(S)$,*

(e) $\partial D^+(S) = H^+(S) \cup S$,

(f) $\partial D(S) = H(S)$,
(g) $I^+[H^+(S)] = I^+[S] - D^+(S)$,
(h) $\operatorname{int} D^+(S) = I^+[S] \cap I^-[D^+(S)]$.

Proof. Exercise. Second exercise: which of (a), (b), $\cdots$, (h) do not require S to be closed? Find "corrected" versions of 5.5 for the remaining cases, when S is not closed, in terms of the following concept.

5.6. DEFINITION. Let S be achronal. The *edge* of S is defined by:

edge $S = \{x|$every neighborhood Q of x contains points y and z and two trips from y to z just one of which meets $S\}$.

Clearly $\bar{S} - S \subset \text{edge } S \subset \bar{S}$, so if we require S to be closed, we have edge $S \subset S$. If edge $S = \varnothing$ we call S *edgeless*. If S is edgeless it must be closed.

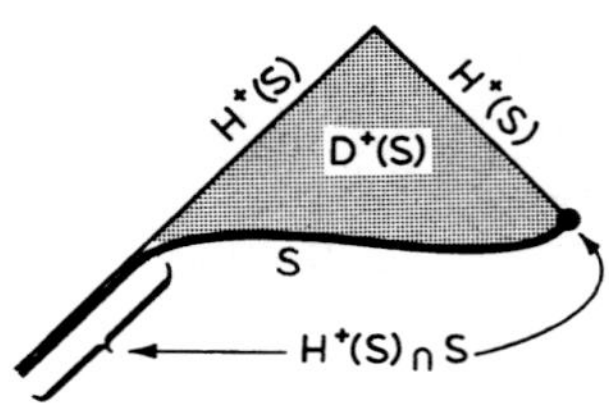

FIG. 36. *An achronal closed set S and its Cauchy horizon $H^+(S)$, with the intersection between the two indicated*

5.7. *Remark.* In the case when S is closed, a slightly different formulation of the definition has been given elsewhere [6], namely: $x \in$ edge S if and only if $x \in S$ and if γ is a trip from y to z containing x, then every neighborhood of γ contains trips from y to z not meeting S. The equivalence of this to 5.6 (S closed and achronal) is evident. (The definition in 5.6 is a corrected version of that given in [9]: a relation $r \ll p \ll q$ on page 191 of that reference should be $p \in \langle r, q \rangle_Q$. Otherwise difficulty arises with examples such as Fig. 23. If S is the line along which strong causality is violated, then, in this example, S should be edgeless.)

The intuitive meaning of edge S is illustrated in Fig. 37. As another example, if S is any spacelike straight line in Minkowski space, then edge $S = S$. In fact,

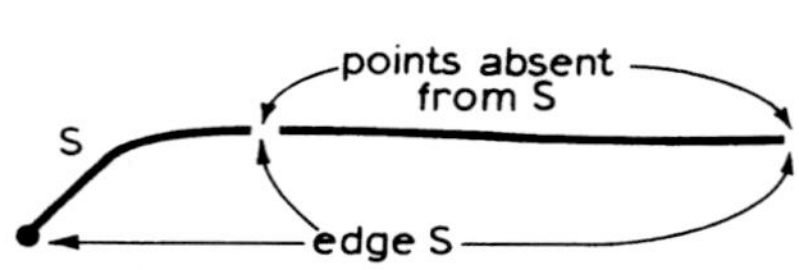

FIG. 37. *An example of an achronal set and its edge, in Minkowski 2-space (here S is not closed)*

edge S is the set of limit points of S not in S, together with the set of points in whose vicinity S fails to be a topological 3-manifold. This is made somewhat more precise in the next proposition, taken in conjunction with 3.17.

5.8. Proposition. *Let $S \subset M$ be achronal. Then $p \notin \text{edge}\, S$ if and only if there is a (connected) open set Q, containing p, such that $S \cap Q$ is an achronal boundary in the space-time manifold Q (where $\varnothing$ is regarded as an achronal boundary in Q).*

Proof. It is clear from the definition in 5.6 and from 3.15, that if $S \cap Q$ is an achronal boundary in the space-time Q, then $p \in Q$ implies $p \notin \text{edge}\, S$. Conversely, suppose $p \notin \text{edge}\, S$. Then there exists a neighborhood P of p such that if $\gamma \subset P$ is a trip from a point $y \in P$ to a point $z \in P$, then any other trip in P from y to z must meet S if and only if γ does. Choose a simple region N, in P, containing p. Choose y and z in N so that $p \in \langle y, z \rangle_N$ (cf. 4.9) and set $Q = \langle y, z \rangle_N$. Then either every trip in N from y to z meets S between y and z, or else every trip in N from y to z misses $S \cap Q$. In the latter case $S \cap Q = \varnothing$ and the required condition is satisfied. In the former case set

$$F_Q = \bigcup_{q \in S \cap Q} \langle q, z \rangle_N .$$

This is clearly a future-set in the space-time Q. Furthermore, any point $x \in Q$ lies on the boundary, in Q, of F_Q if and only if it lies on S (as follows from the achronality of S, by consideration of trips from y to z via x). Thus $S \cap Q$ is an achronal boundary in the space-time Q.

5.9. Corollary. *Any achronal boundary in M is edgeless.*

5.10. Proposition. *If S is achronal,* edge S *is closed.*

Proof. Immediate from 5.8.

5.11. Proposition. *If S is achronal, then:*

(a) $I^+[\text{edge}\, S] \cap D^+(S) = \varnothing$,

(b) $\text{edge}\, S = \text{edge}\, H^+(S)$.

Proof. Exercise.

5.12. Theorem [4], [9]. *Let S be achronal. Then every point of $H^+(S) -$ edge S is the future endpoint of a null geodesic on $H^+(S)$ which is either past-endless or else has past endpoint on* edge S.

Proof. The idea is to use 3.19. For this we need a suitable future set. Define $W = I^+[S] - D^+(S)$. In fact we have $W = I^+[H^+(S)]$ by 5.5(g) (which actually does not require S to be closed) showing that W is a future set. However, it will be better to think of W in a different way. We have: $x \in W$ if and only if there is both a past-endless trip α terminating at x and not meeting S and another trip β from a point of S to x (see Fig. 38). It is readily seen from this that W is open and that $I^+[W] \subset W$, S being achronal, so by 3.6, W is a future set: $I^+[W] = W$. Now $H^+(S)$ is a part of the achronal boundary ∂W. (Actually $H^+(S) = \partial W \cap D^+(S)$.) The remaining part of ∂W is $\partial I^+[S] - S$. In fact, since $\partial W \cap W = \varnothing$ it must be that for $x \in \partial W$ either the α-trip or the β-trip defined above fails to exist. If $x \in \partial I^+[S] - S$, then the α-trip exists but not the β-trip. If $x \in H^+(S) - S$, the β-trip exists but not the α-trip. If $x \in S$, the β-trip becomes degenerate and the α-trip fails to exist.

Now suppose $p \in H^+(S) -$ edge S. We can choose a simple region $Q \ni p$ so that $\partial I^+[S] \cap \bar{Q} = S \cap \bar{Q}$. If $p \in H^+(S) - S$, we do this by choosing $\bar{Q}$ inside the

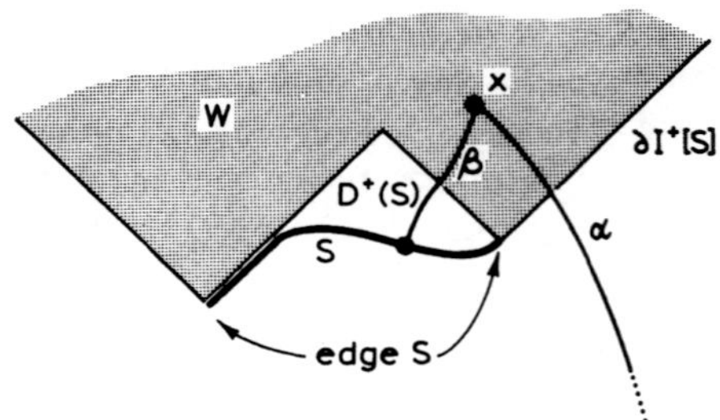

FIG. 38. *The future set W consists of all points which lie both on a past-endless trip not meeting the achronal set S and also on a trip with past endpoint on S. The achronal boundary ∂W consists partly of $\partial I^+[S] - S$ and partly of $H^+(S)$*

future of a β-trip from a point of S to p. If $p \in S$ − edge S, we do this by taking $\bar{Q}$ within a set $\langle y, z \rangle_N$ (cf. proof of the proposition in 5.8), inside which every trip from y to z meets S. The implication is that every point of $\bar{Q} \cap \overline{I^+[S]}$ is the future endpoint of a possibly degenerate β-trip. But any $x \in I^+(p)$ must lie in W and so is the future endpoint of an α-trip which must meet $\partial Q \cap \overline{I^+[S]}$ in some point q. Now q is the future endpoint of a β-trip also, so $q \in W \cap \partial Q \subset W - Q$. Thus 3.19 is satisfied and we have p as the future endpoint of a null geodesic η on $H^+(S)$. We could repeat the argument at any past endpoint of η. So (by the achronality of $H^+(S)$) we can extend η into the past along $\partial H^+(S)$ either indefinitely or until it meets edge S.

5.13. DEFINITION. A *Cauchy hypersurface for M* (sometimes called a global Cauchy hypersurface) is an achronal set S for which $D(S) = M$.

5.14. PROPOSITION [9]. *If S is achronal and intersects every endless null geodesic in M in a nonempty compact set, then S is a Cauchy hypersurface for M.*

Proof. If $D(S) \neq M$, then (by 5.5(f)) either $H^+(S)$ or $H^-(S)$ must be nonempty. Suppose $H^+(S) \neq \varnothing$. Then there is a null geodesic on $H^+(S)$ whose maximal extension γ must, by hypothesis, intersect S in a nonempty compact set. We can follow $\gamma \cap S$ into the past along γ until we reach edge S (since $\gamma \cap S$ is compact). We obtain the desired contradiction by showing that edge S must be empty. This will follow from 5.9 if we can show that $S = \partial I^+[S]$. Now $S \subset \partial I^+[S]$ since S is achronal. Suppose $p \in \partial I^+[S]$ but $p \notin S$. Choose an endless null geodesic η through p. Since $\eta \cap S$ is closed (being compact), a point q exists on η between p and $\eta \cap S$ and we must have $q \in \partial I^+[S]$. Any null geodesic ζ through q with a different direction from that of η can meet $\partial I^+[S]$ only at q. But this contradicts the hypothesis, since $q \notin S$.

5.15. *Remark.* If S is both achronal and closed, then we can replace the condition "in a nonempty compact set" in 5.14 by some weaker condition. But if S is not assumed to be closed, this condition, or something like it, is necessary. For example, if S is the union of the regions $t = 1, x^2 + y^2 + z^2 \geqq 1$ and $0 < t \leqq 1, t^2 - x^2 - y^2 - z^2 = 0$ of Minkowski space, then every endless null geodesic meets S, but not every endless trip. Hence $D(S) \neq M$. On the other hand, if S is smooth and spacelike everywhere, then we need not assume it is closed in order to deduce that it is a Cauchy hypersurface merely from the fact that it meets every endless null geodesic (exercise).

5.16. PROPOSITION. *If S is achronal and $x \in D^+(S) - H^+(S)$, then every past-endless causal trip with future endpoint x must intersect $S - H^+(S) -$ edge S and must contain a point in $I^-[S]$.*

Proof. If $x \in S$ the conclusion is trivial. So assume $x \in \operatorname{int} D^+(S) = D^+(S) - H^+(S) - S$ (cf. 5.5(e); but S need not be closed here). Then there is a point $y_1 \in I^+(x) \cap D^+(S)$. Let γ be a past-endless causal trip with future endpoint x. Cover γ by a locally finite system of simple regions $N_1, N_2, \cdots$. Refer, now, to Fig. 39. We have $x = x_1 \in N_{i_1}$ for some i_1, and we can choose y_1 to be in N_{i_1}.

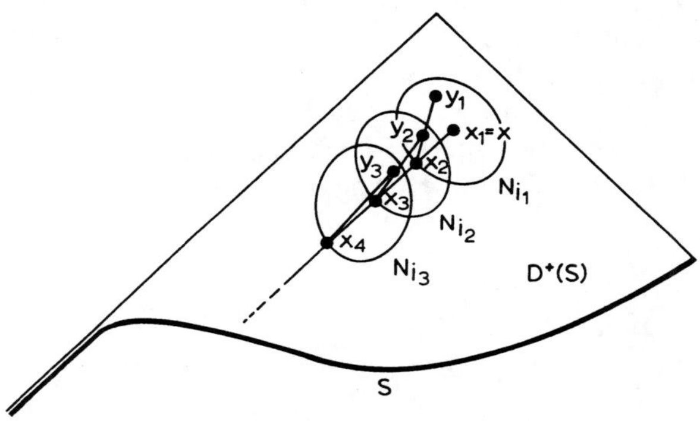

FIG. 39. *Diagram for the proof of* 5.16

Let x_2 be the past endpoint of the connected component of $\gamma \cap \overline{N}_{i_1}$ to x_1, so $x_2 \in \partial N_{i_1}$ with $x_2 \prec x_1$. Thus $x_2 \ll y_1$. We have $x_2 \in N_{i_2}$ for some $i_2 \neq i_1$. Choose $y_2 \in N_{i_2}$ with $x_2 \ll y_2 \ll y_1$. Let $x_3 \in \partial N_{i_2}$ be the past endpoint of the connected component of $\gamma \cap \overline{N}_{i_2}$ to x_2. Then $x_3 \in N_{i_3}$ for some $i_3 \neq i_2$. Continuing indefinitely in this way we obtain $\cdots \ll y_3 \ll y_2 \ll y_1$ with $y_1 \in D^+(S)$ and with $y_{r+1}y_r$ future-timelike in N_r, $r = 1, 2, \cdots$. Since no single segment of the causal trip γ can enter and leave one N_i more than once, and since the $\{N_i\}$ is a locally finite system, it follows that the x_i's must proceed indefinitely into the past along γ (i.e., with no point $x' \in \gamma$ preceding all of the x_i's). Hence the y_i's must lie in infinitely many of the N_i's. It follows that $\cdots \cup y_4y_3 \cup y_3y_2 \cup y_2y_1$ constitutes a genuine past-endless trip η (and not a "bad trip," cf. 2.1). Since $y_1 \in D^+(S)$, η must meet S at z, say, with z on the segment y_ky_{k-1}, say. We have $x_k \ll z$, so $x_k \notin D^+(S)$. Thus some point w of γ lies on $\partial D^+(S)$. The cases $w \in H^+(S)$ or $w \in$ edge S cannot occur, since, by 5.11(a), $w \prec x$ would imply $w \in H^+(S) \cup \sim D^+(S)$. (See 5.5 and its extension when S is not closed.) Thus $w \in S - H^+(S) -$ edge S. Also, $x_k \in I^-[S]$.

5.17. PROPOSITION. *Let S be achronal. If $y \in \operatorname{int} D^+(S)$, then $J^-(y) \cap I^+[S] = J^-(y) \cap \operatorname{int} D^+(S)$ and $J^-(y) \cap J^+[S] = J^-(y) \cap D^+(S)$.*

Proof. Exercise.

5.18. PROPOSITION [4]. *If $S \subset M$ is achronal and $p \in \operatorname{int} D^+(S)$, then M is strongly causal at p.*

Proof. Suppose, first, that some point of $D^+(S)$ lies on a closed trip η. Such an η is past-endless and so must meet S in some point w. But this gives $w \ll w$ contradicting the achronality of S. Thus $D^+(S) \cap V = \varnothing$ (cf. 4.26), so $\operatorname{int} D^+(S) \cap \bar{V} = \varnothing$. Now suppose strong causality fails at some point $p \in \operatorname{int} D^+(S)$. By 4.31 there must be an endless null geodesic γ through p with the property that if $q \in \gamma$ with $q \prec p$, $q \neq p$, then every $x \in I^+(q)$ and $y \in I^-(p)$ must satisfy $y \ll x$. (This is because all cases (a), (b), (c), (d) of 4.31 require $p \in \bar{V}$, leaving us only with case (e).) By 5.16, γ must contain some point $q \in I^-[S]$. Since $I^-[S]$ and $\operatorname{int} D^+(S)$ are both open, we can find $x \in I^+(q) \cap I^-[S]$ and $y \in I^-(p) \cap \operatorname{int} D^+(S)$. Then $y \ll x$. But also a trip exists from a point of S to y ($y \in D^+(S)$) and another trip exists from x to a point of S ($x \in I^-[S]$). The resulting violation of the achronality of S yields the required contradiction.

5.19. *Remark.* Examples can be constructed in which strong causality fails on $\partial \operatorname{int} D^+(S)$. In Fig. 40 strong causality fails on a part of S; in Fig. 41 it fails on a part of $H^+(S)$. In each case the space-time is the same as that of Fig. 23.

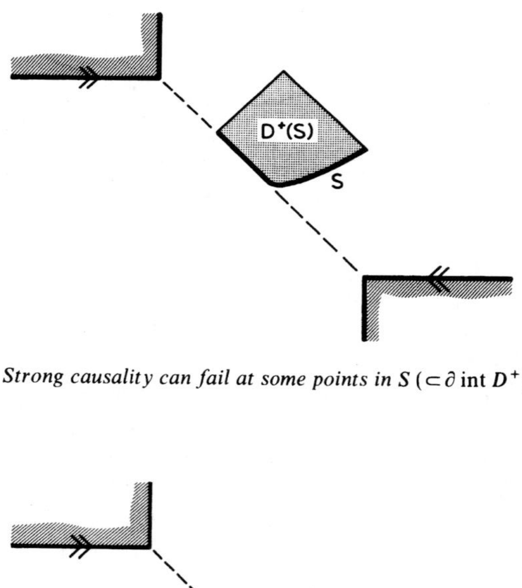

FIG. 40. *Strong causality can fail at some points in S ($\subset \partial \operatorname{int} D^+(S)$ here)*

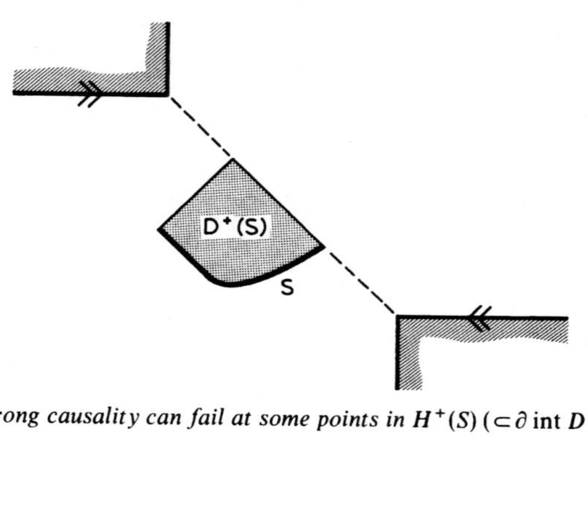

FIG. 41. *Strong causality can fail at some points in $H^+(S)$ ($\subset \partial \operatorname{int} D^+(S)$ here)*

5.20. PROPOSITION [4]. *If S is achronal and $x \in \operatorname{int} D^+(S)$, then $J^-(x) \cap J^+[S]$ is compact.*

Proof. Set $A = J^-(x) \cap J^+[S]$, with $x \in \operatorname{int} D^+(S)$. Suppose A is not compact. Then there is a sequence of points $a_0, a_1, a_2, \cdots \in A$ with no accumulation point in A. The idea is to use this sequence to construct a past-endless trip γ with future endpoint in $D^+(S)$ but which does not meet S, thus supplying a contradiction. Cover A with a locally finite system of simple regions $\{N_i\}$. Refer, now, to Fig. 42. Suppose $x = x_0 \in N_{i_0}$. We can choose $y_0 \in I^+(x) \cap D^+(S) \cap N_{i_0}$. Now $a_i \in J^-(x)$,

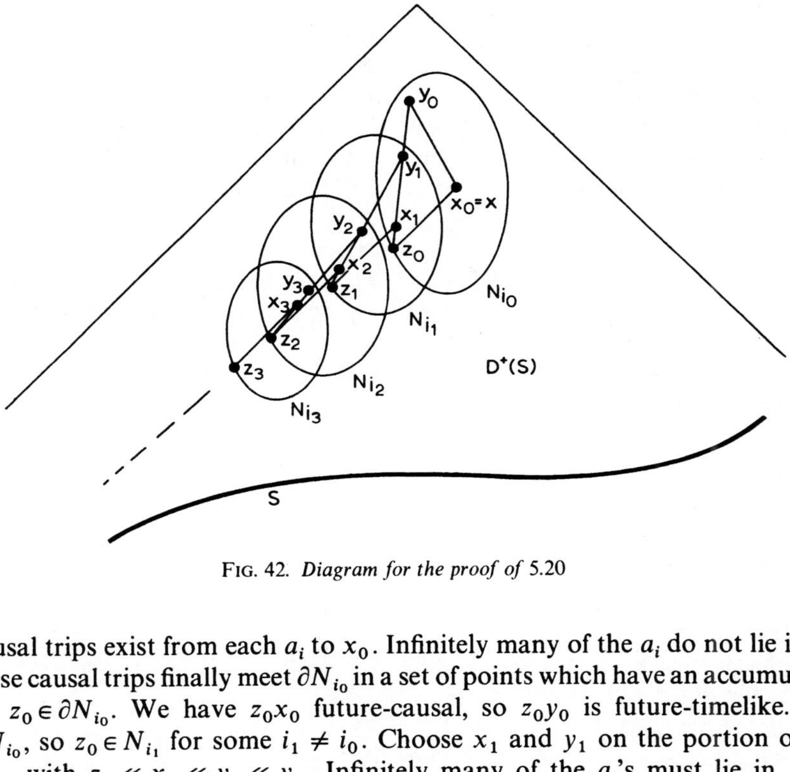

FIG. 42. *Diagram for the proof of* 5.20

so causal trips exist from each a_i to x_0. Infinitely many of the a_i do not lie in N_{i_0}, so these causal trips finally meet ∂N_{i_0} in a set of points which have an accumulation point $z_0 \in \partial N_{i_0}$. We have $z_0 x_0$ future-causal, so $z_0 y_0$ is future-timelike. Now $z_0 \notin N_{i_0}$, so $z_0 \in N_{i_1}$ for some $i_1 \neq i_0$. Choose x_1 and y_1 on the portion of $z_0 y_0$ in N_{i_1}, with $z_0 \ll x_1 \ll y_1 \ll y_0$. Infinitely many of the a_i's must lie in $I^-(x_1)$ (since $I^-(x_1)$ is open), so again there must be a point $z_1 \in \partial N_{i_1}$ which is an accumulation point of final intersections of causal trips from a_i's to x_1. We have $z_1 x_1$ future-causal, so $z_1 y_1$ is future-timelike. Proceeding exactly as before we construct x_2, y_2 with $z_1 \ll x_2 \ll y_2 \ll y_1$ and then x_3, y_3, etc. This yields a sequence of points $y_0, y_1, y_2, \cdots$ with $\cdots \ll y_2 \ll y_1 \ll y_0$. The union of segments: $\cdots \cup y_3 y_2 \cup y_2 y_1 \cup y_1 y_0$ constitutes a genuine past-endless trip γ since (by the locally finite nature of $\{N_i\}$) the y_i's cannot accumulate in a single N_i to produce a "bad trip." Now γ cannot meet S since otherwise we should have some $y_i \in I^-[S]$ which would be inconsistent with $a_j \ll y_i$ and $a_j \in J^+[S]$ (whence $y_i \in I^+[S]$) because S is achronal.

5.21. PROPOSITION. *If S is achronal and $y \in \operatorname{int} D^+(S)$, then $\overline{I^-(y) \cup D^+(S)} = J^-(y) \cap D^+(S)$.*

Proof. We have $I^-(y) \cap D^+(S) \subset J^-(y) \cap D^+(S) \subset \overline{I^-(y) \cap D^+(S)}$. But $J^-(y) \cap D^+(S) = J^-(y) \cap J^+[S]$, which is compact (by 5.20) and therefore closed. The result follows.

5.22. PROPOSITION. *If S is achronal, then strong causality holds throughout* $\operatorname{int} D(S)$.

Proof. Exercise.

5.23. PROPOSITION. *If S is achronal and $u, v \in \operatorname{int} D(S)$, then $J^+(u) \cap J^-(v)$ is compact.*

Proof. Exercise.

5.24. DEFINITION. A space-time M is said to be *globally hyperbolic* if and only if M is strongly causal and every set $J^+(u) \cap J^-(v)$, with $u, v \in M$, is compact. (A slightly different, but equivalent definition is more usual, stating the compactness of the space of causal curves connecting u to v; see 6.8.)

5.25. THEOREM [30]. *M is globally hyperbolic if and only if a Cauchy hypersurface exists for M.*

Proof. If $M = D(S)$ for some achronal S, then $D(S) = \operatorname{int} D(S)$, so global hyperbolicity follows from 5.22 and 5.23. For the converse, see Geroch [30].

5.26. THEOREM [30]. *If a Cauchy hypersurface S exists for M, then M is homeomorphic to $\mathbb{R} \times S$. Furthermore, if $f\colon \mathbb{R} \times S \to M$ is the homeomorphism, we can arrange it so that $f(t, S)$ is a Cauchy hypersurface for each t and $f(\mathbb{R}, s)$ is a timelike curve for each $s \in S$.*

Proof. See Geroch [30].

SECTION 6

The Space of Causal Curves

6.1. DEFINITION. Let K denote the subset of M consisting of all points at which M is strongly causal. By 4.13, K is open. Let $\mathscr{C}$ denote the set of all causal curves lying in K (cf. 2.25). Let $\mathscr{K}$ denote the set of all causal trips in K and $\mathscr{T}$ denote the set of all trips in K. We have

$$\mathscr{T} \subset \mathscr{K} \subset \mathscr{C}.$$

Let C be a subset of K and let A and B be subsets of C. Define

$$\mathscr{C}_C(A, B) = \{\gamma | \gamma \text{ is a causal curve in } C \text{ from a point of } A \text{ to a point of } B\}.$$

The notation $\mathscr{C}(A, B)$ will also be used for the above, but with "in C" deleted (i.e., with K replacing C).

I shall be interested in these sets particularly when C is compact and when A and B are closed. The idea then will be to topologize $\mathscr{C}$ in a natural way, so that any such $\mathscr{C}_C(A, B)$ becomes compact. A length function $l:\mathscr{C}_C(A, B) \to \mathbb{R}$ ("proper time") will be defined and shown to be upper semicontinuous. Thus the compactness will imply that l attains a maximum[1] value on $\mathscr{C}_C(A, B)$. Under suitable circumstances this maximum is attained by a geodesic without conjugate points (cf. 1.18). This fact forms the basis of most of the "singularity theorems" referred to in the introduction.

6.2. DEFINITION. We topologize $\mathscr{C}$ by taking as a base for open sets in $\mathscr{C}$ the sets of the form $\mathscr{C}_R(P, Q)$ where P, Q and R are open sets in N with $P, Q \subset R$. It is clear that the sets $\mathscr{C}_R(P, Q)$ do form a base for a topology since if $\gamma \in \mathscr{C}_R(P, Q)$ and $\gamma \in \mathscr{C}_{R'}(P', Q')$, then $\gamma \in \mathscr{C}_{R''}(P'', Q'') \subset \mathscr{C}_R(P, Q) \cap \mathscr{C}_{R'}(P', Q')$, where $R'' = R \cap R'$, $P'' = P \cap P'$, $Q'' = Q \cap Q'$.

6.3. *Remark.* This simple definition of a topology on $\mathscr{C}$ has been used by Geroch [30]. It agrees with an intuitive notion of "C^0-topology" on curves, whereby no heed is paid to smoothness or to the directions of tangents (cf. Fig. 43). This is necessary if, for example, we desire causal curves to arise as limits of trips or of causal trips, and so that $\mathscr{C}$ can be locally compact [35]. However we are only at liberty to use this definition because we have excluded the region of strong

[1] A feature of hyperbolic normal manifolds is that lengths (for timelike curves) are locally *maximized* by geodesics rather than minimized (which is the familiar situation for positive definite spaces). This is related to the familiar "clock paradox": accelerated observers generally experience shorter time intervals than unaccelerated ones.

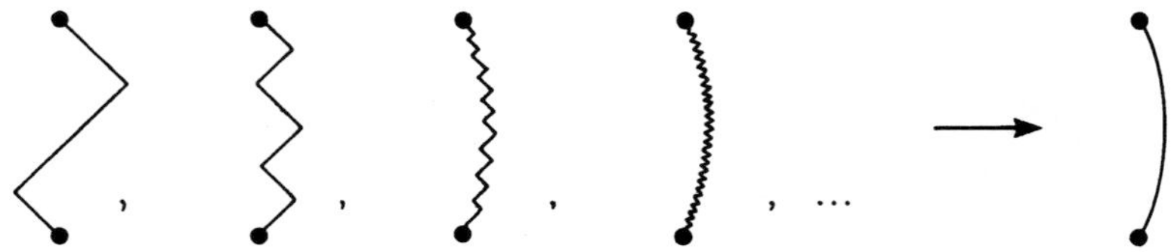

FIG. 43. *The topology assigned to $\mathscr{C}$ allows smooth curves to arise, for example, as a limit of a sequence of (causal) trips. The direction of the tangent vectors is of no consequence and need not approach a limit*

causality failure (or at least the region of closed causal trips) from our consideration. If we had not done so, a slightly more sophisticated definition would have been required [20].

6.4. PROPOSITION. *$\mathscr{K}$ is dense in $\mathscr{C}$ and $\mathscr{T}$ is dense in $\mathscr{K}$.*

Proof. Let $\gamma \in \mathscr{C}$ and let $\mathscr{R} = \mathscr{C}_R(P, Q)$ be a neighborhood of γ in $\mathscr{C}$ (P, Q, R being open in M). We can cover γ by simple regions contained in R and use these to obtain a causal trip η contained in $\mathscr{R}$ (cf. the definition in 2.25 of a causal curve). The construction is indicated in Fig. 44 and is straightforward. If $\gamma' \in \mathscr{K}$ and $\mathscr{R}' = \mathscr{C}_{R'}(P', Q')$ is a neighborhood of γ', we can obtain a trip $\eta' \in \mathscr{R}'$ as follows.

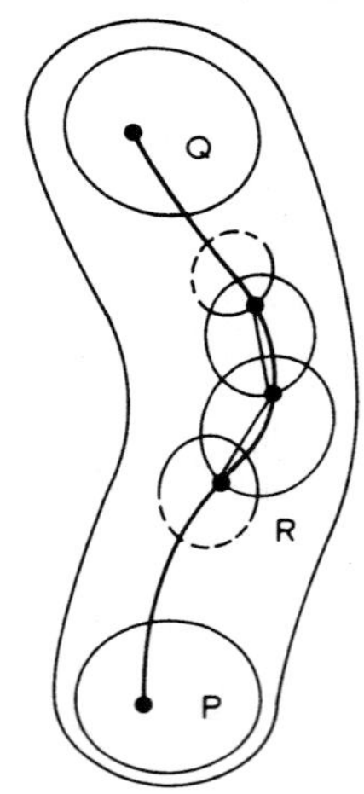

FIG. 44. *How to approximate a causal curve by a causal trip, as required for* 6.4

Let γ' have past endpoint p and future endpoint q. Choose $r \in Q'$ with $q \ll_{R'} r$. We have $p \prec_{R'} q$, whence $p \ll_R r$, so we can take η' from p to r in Q'.

6.5. THEOREM. *If C is a compact subset of K and A and B are closed subsets of C, then $\mathscr{C}_C(A, B)$ is compact.*

Proof. I do not have a nice simple argument, but I feel sure that one must exist (exercise: find one!). The argument I do have is rather untidy so I give it somewhat informally. The idea is to show that any infinite sequence of causal curves $\gamma_i \in \mathscr{C}_C(A, B)$ has an accumulation curve $\gamma \in \mathscr{C}_C(A, B)$. Since A is compact, an accumulation point p of the past endpoints of the γ_i's exists in A. Select a subsequence of the γ_i's whose past endpoints converge on p. B is compact so an accumulation future endpoint q exists. Select a subsequence with future endpoints converging on this also. Cover C with a finite number of local causality neighborhoods. One of these, N_0, contains $p = p_0$. Select a subsequence of the resulting γ_i's converging also on an accumulation point $p_1 \in \partial N_0$ of points at which the γ_i's leave N_0. Then p_1 lies in another member of the covering, say N_1. Repeat the argument to obtain p_2, p_3, etc. We end up with a finite sequence $p = p_0, p_1, p_2, p_3, \cdots, p_k = q$ of limit points of points on our resulting subsequence of γ_i's, with $p_{i-1} \prec p_i$, $i = 1, 2, \cdots, k$, each consecutive pair lying in the closure of a local causality neighborhood N_{i-1}. To construct a causal curve from p_{i-1} to p_i we now proceed as follows. We may suppose that N_0 has been chosen small enough so that the hypersurfaces $t =$ const., in a Minkowski normal coordinate system with origin p_0, are spacelike. Let the value of the t-coordinate of p_1 be ε. Consider the intersection of the γ_i subsequence with the hypersurface $t = \frac{1}{2}\varepsilon$. We obtain an accumulation point $r_{0.1}$ and a new subsequence of γ_i's converging on this also. Repeat, with p_2 in place of p_1 and p_1 in place of p_0 to obtain $r_{1.1}$ between p_1 and p_2 and a new subsequence converging on this also. Repeat for $p_3, p_4, \cdots, p_k = q$. Then return to N_0 and repeat the construction with $t = \frac{1}{4}\varepsilon$ and then with $t = \frac{3}{4}\varepsilon$, to obtain $r_{0.01}$ and $r_{0.11}$, respectively. And so on. The construction gives us a point r_α for each real number between 0 and k whose binary expansion α terminates. We have $r_\alpha \prec r_\beta$ if $\alpha \leqq \beta$. (The r_α's constitute a causal chain [18].) Each r_α is an accumulation point of the curves of some subsequence of the γ_i's and hence of the curves of the original sequence. The closure $\overline{\bigcup_\alpha \{r_\alpha\}}$ is the desired accumulation causal curve.

6.6. Corollary. *Let S be achronal and suppose strong causality holds at each point of S. Let $x \in \operatorname{int} D^+(S)$ and $y, z \in \operatorname{int} D(S)$. Then $\mathscr{C}(S, \{x\})$ and $\mathscr{C}(\{y\}, \{z\})$ are compact.*

Proof. Result follows from 5.18, 5.20, 5.22, 5.23.

6.7. *Remark.* If we replace "int $D^+(S)$" by "$D^+(S)$," or "int $D(S)$" by "$D(S)$," the result becomes untrue. (Exercise: find counterexamples.)

6.8. *Remark.* By virtue of 6.6, we see (cf. 5.24, 5.25) that in any globally hyperbolic space-time:

$$\mathscr{C}(\{x\}, \{y\}) \quad \text{is compact for all} \quad x, y \in M.$$

In fact, had we given a suitable definition of the topology of $\mathscr{C}$ for regions of M where there are closed causal trips, this property gives essentially Leray's original formulation of global hyperbolicity [20].

SECTION 7

Geodesics as Maximal Curves

7.1. Definition. Let γ be a causal trip. Define the *length* (i.e., "proper time") of γ to be:

$$l(\gamma) = \sum_{i=1}^{k} \{\Phi(p_{i-1}, p_i)\}^{1/2},$$

where successive segments of γ are $p_0p_1, p_1p_2, \cdots, p_{k-1}p_k$ (each segment $p_{i-1}p_i$ for definiteness, lying within a simple region N_i) and where Φ is the world function defined in 2.13. (We have $\Phi(p_{i-1}, p_i) \geqq 0$ since $p_{i-1}p_i$ is causal.) This definition simply assigns the obvious meaning of length, according to the space-time metric, to any causal trip. Clearly $l(\gamma) > 0$ unless γ consists entirely of null segments.

7.2. Proposition. *Let N be a simple region and let $p, q \in N$ with pq future-causal. Then if η is the causal trip pq and η' is any other causal trip in N from p to q, we have $l(\eta) > l(\eta')$.*

Proof. If pq is null, the result is obvious (and vacuous) from 2.19. Let pq be timelike and choose Minkowski normal coordinates (t, x, y, z) for N, with origin at some point r, in N, lying to the past of p along the extension of pq. Choose new coordinates for the region $\hat{N}$ given by $t > (x^2 + y^2 + z^2)^{1/2}$ as follows:

$$T = (t^2 - x^2 - y^2 - z^2)^{1/2},$$

$$X^1 = \frac{x}{t}, \quad X^2 = \frac{y}{t}, \quad X^3 = \frac{z}{t}.$$

Since the curves $X^1, X^2, X^3 = \text{const.}$ are timelike geodesics through r, and $T = \text{const.}$ are spacelike hypersurfaces orthogonal to these (cf. 2.15), where $T\ (= \{\Phi(r,\ )\}^{1/2})$ measures the length (i.e., proper time) on the geodesic from r, we have what is known as a *synchronous coordinate system* for N (i.e., a Gaussian normal coordinate system in which the geodesics are timelike, being orthogonal to a system of spacelike coordinate hypersurfaces). The metric therefore has the form

$$ds^2 = dT^2 - \gamma_{\alpha\beta}\, dX^\alpha\, dX^\beta, \qquad \alpha, \beta = 1, 2, 3,$$

where at each point of $\hat{N}$, $(\gamma_{\alpha\beta})$ is a positive definite matrix. Since

$$l(\eta') = \int_{T_0}^{T_1} \left(1 - \gamma_{\alpha\beta} \frac{dX^\alpha}{dT} \frac{dX^\beta}{dT}\right)^{1/2} dT,$$

where T_0 and T_1 are the T-coordinates of p and q, respectively, it is clear that the maximum is uniquely attained when the X^α-coordinates are constant, this giving the geodesic η.

7.3. *Remark.* Clearly the proof in 7.2 would work equally well for any "rectifiable" causal curve η' (in the sense that the length integral exists) from p to q. However, *every* causal curve is "rectifiable" as the following definition shows.

7.4. DEFINITION. Let $p \prec q$ and let γ be a causal curve from p to q. Let $\xi = \{x_i\}$ denote a finite sequence of points along γ, beginning at $p = x_0$ and terminating at $q = x_k$, such that any consecutive pair x_i, x_{i+1} are contained in a simple region N_i which also contains the portion of γ from x_i to x_{i+1}. We have $x_i \prec x_{i+1}$ so x_ix_{i+1} is future-causal. The symbol γ_ξ denotes the causal trip $x_0x_1 \cup x_1x_2 \cup \cdots \cup x_{k-1}x_k$. Let Ξ be the set of all such allowable sequences ξ. The notations $\xi \subset \xi'$ and $\xi \cup \xi' = \xi''$ have their obvious meanings. Clearly

$$l(\gamma_{\xi'}) \leqq l(\gamma_\xi) \quad \text{if} \quad \xi \subset \xi'$$

by repeated application of 7.2. Also, given $\xi, \xi' \in \Xi$ we have

$$l(\gamma_{\xi''}) \leqq \min(l(\gamma_\xi), l(\gamma_{\xi'})),$$

where $\xi'' = \xi \cup \xi'$. Finally, define $l:\mathscr{C}(\{p\}, \{q\}) \to \mathbb{R}$ by

$$l(\gamma) \equiv \inf_{\xi \in \Xi} \{l(\gamma_\xi)\}.$$

The infimum clearly exists since $l(\gamma) \geqq 0$ and so assigns a meaning to the concept of the length of any causal curve with two endpoints. This definition also extends to the whole of $\mathscr{C}$, i.e., to past- or future-endless causal curves if we allow the value ∞ for $l(\gamma)$. Thus, $l:\mathscr{C} \to \mathbb{R} \cup \{\infty\}$. I shall only be concerned with causal curves having both past and future endpoints, however. Then we can regard l as a map $l:\mathscr{C}_C(A, B) \to \mathbb{R}$, for any $A \subset C$, $B \subset C$, $C \subset K$.

7.5. THEOREM. *The map $l:\mathscr{C}_C(A, B) \to \mathbb{R}$ is upper semi-continuous.*

Proof. We have to show that l^{-1} applied to any set in $\mathbb{R}$ of the form $(-\infty, a)$ is open in $\mathscr{C}_C(A, B)$ (cf. [35]). This will follow if we can show that given any causal curve $\gamma \subset C$, from $p \in A$ to $q \in B$, satisfying $l(\gamma) < a$, there is a neighborhood $\mathscr{R}$ of γ in $\mathscr{C}$ such that any $\gamma' \in \mathscr{R}$ also satisfies $l(\gamma') < a$. Suppose $l(\gamma) = b < a$. Choose $\xi \in \Xi$ (cf. 7.4) such that $l(\gamma_\xi) < b + \frac{1}{2}(a - b)$. Suppose that the x_i are chosen close enough to each other along γ so that each consecutive pair x_i, x_{i+1}, $i = 0, \cdots, k - 1$ is contained in some local causality neighborhood L_i, $\bar{L}_i \subset N_i$ (cf. 4.11) and, furthermore, so that L_i intersects L_j only if $j = i \pm 1$ (see Fig. 45). Since the length of a geodesic in N_i is a continuous function of its endpoints (cf. 2.14) it follows that we can choose a local causality neighborhood U_i of each x_i (with $U_i \subset L_i$, $U_{i+1} \subset L_i$, $i = 0, 1, \cdots, k - 1$) small enough that any causal geodesic from a point of U_i to a point of U_{i+1} must differ in length from $l(x_ix_{i+1})$ by less than $|a - b|/2k$.

Set

$$V_i = \bigcup_{\substack{y \in U_i \\ z \in U_{i+1}}} \langle y, z \rangle.$$

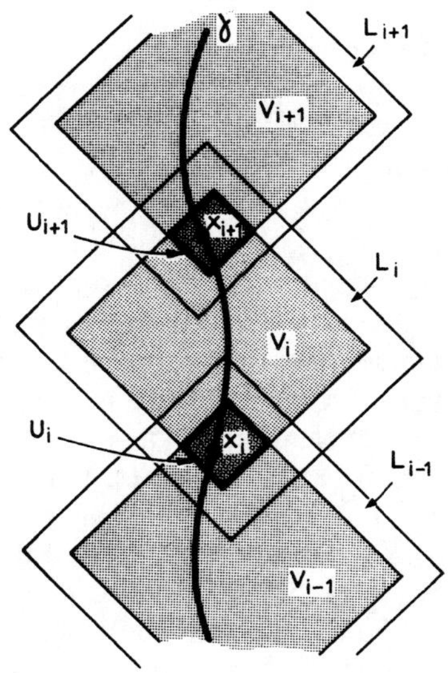

FIG. 45. *Diagram for the proof of* 7.5

Then $V_i \subset L_i$, by the causal convexity of L_i (cf. 4.4), whence V_i intersects V_j only if $j = i \pm 1$.

Define $P = V_0$, $Q = V_{k-1}$ and $R = \bigcup_i V_i$. Suppose $\gamma' \in \mathscr{C}_R(P, Q)$. Then γ' threads through the V_i consecutively. Furthermore, γ' must pass through each U_i. This follows from the causal convexity of U_i since γ' meets $V_{i-1} \cap V_i$ in a point which must (from the definition of the V's) lie in some p, q with $p, q \in U_i$. Thus, γ' contains points $x'_0, x'_1, \cdots, x'_k$ with $x_i \in U_i$, so that the causal trip $\eta = x'_0x'_1 \cup x'_1x'_2 \cup \cdots \cup x'_{k-1}x_k$ satisfies $l(\eta) < b + \frac{1}{2}(a - b) + kx(2k)^{-1}(a - b) = a$. Hence $l(\gamma') < a$ as required.

7.6. *Remark.* Though upper semi-continuous, the map l cannot be continuous. The illustration in Fig. 43 makes this clear. In an extreme case we could envisage a timelike curve of well-defined nonzero length (e.g., a timelike geodesic) arising as a limit of a sequence of causal trips each of whose segments is null so that the total length of each is zero.

7.7. COROLLARY.[1] *If A and B are closed subsets of a compact set C, throughout which strong causality holds, then there is a causal curve in C from a point of A to a point of B which maximizes the lengths of such curves.*

Proof. This follows at once from 6.5 and 7.5 (cf. [35]).

7.8. PROPOSITION. *Let A, B and C be as in 7.7 and let $\gamma \in \mathscr{C}_C(A, B)$ maximize $l(\gamma)$. Then $\gamma \subset$ int C implies that γ is a causal geodesic (possibly degenerate).*

Proof. If $\gamma \subset$ int C, we can cover γ with a system of simple regions contained in C. The fact that the intersection of γ with each simple region must be a geodesic, follows from 7.2 and 7.4. Hence γ itself must be a geodesic.

[1] The use of *trips* in the various developments leading up to and establishing this result arose from a discussion with Robert Geroch.

7.9. *Remark.* We have seen in 7.2 that a causal geodesic is locally a curve of maximum length; also that under suitable circumstances a curve of maximal length is a causal geodesic. However, it is not always true that a given causal geodesic from p to q is the curve of maximal length from p to q, or even that such a maximal curve exists in all cases. For example, we can refer to Fig. 7 ("anti-deSitter space") in which two points a and x satisfy $a \ll x$, but no geodesic connects them. A timelike geodesic connects c to b, on the other hand, but one can see that this does not actually maximize the length of causal curves from c to b. For there are many geodesics from a to b. If we choose one of these which does not continue our choice of geodesic from c to a, we obtain a trip from c to b with a "joint" at a. By "cutting the corner" at the joint, to produce a trip with three segments, we clearly obtain a trip of greater length than that of the original geodesic from c to b. The crucial fact here is that this geodesic from c to b contains pairs of *conjugate points.*[2] This concept was briefly introduced in 1.18: if γ is a geodesic and V is a nontrivial Jacobi field defined on γ which vanishes at two distinct points p and q on γ, then p and q are called conjugate points on γ.

A rough intuitive picture of why a causal geodesic is not maximal if it contains a pair of conjugate points (not at its endpoints) is obtainable from Fig. 46. Here the conjugate points p and q occur between a and b on γ. We can crudely imagine a

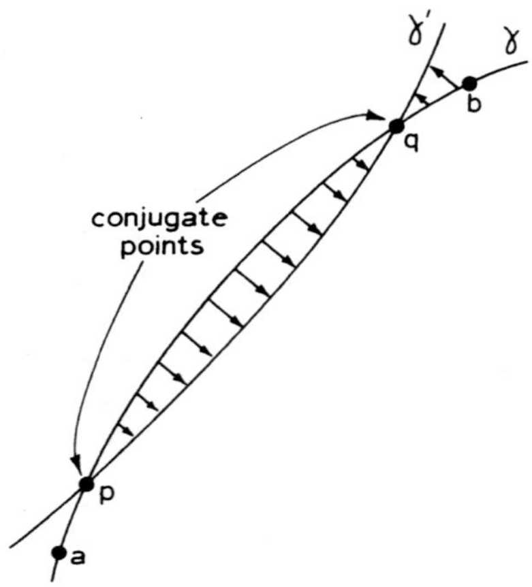

FIG. 46. *A pair of conjugate points on a (causal) geodesic γ may be thought of as a pair of intersection points of γ with a "neighboring geodesic" γ' to γ*

[2] A discussion of the physical significance of conjugate points on a timelike geodesic in relation to the "clock paradox" has been given by Boyer [34]. For example, the world-line of the earth as it revolves once around the sun, from one event p to a later event q, with the same spatial location as p, is a geodesic. Nevertheless the time-interval experienced on the earth is *less* than that which it would have been, had the earth remained at the same spatial location from p to q, this being *not* a geodesic from p to q. The reason for this is that the earth in orbit encounters a conjugate point to p when it is half way from p to q (at the far side of the sun). The maximum time from p to q is in fact attained by a geodesic (without conjugate points) representing the free-fall outwards from p, returning inwards towards the sun to q.

"neighboring geodesic" γ' to connect p to q, having length essentially the same as that of the portion of γ from p to q. Then if we proceed from a to p along γ, from p to q along γ', and finally from q to b along γ, we obtain a causal trip from a to b whose length is essentially the same as that of γ. But this causal trip has two "joints," so we can "cut the corners" to obtain a new trip from a to b of length greater than that of γ. However, this argument is very crude as it stands (and is even fallacious to some extent). It is not easy to make it rigorous by "putting in ε's." In Fig. 47 the situation is given in a little more detail near q,

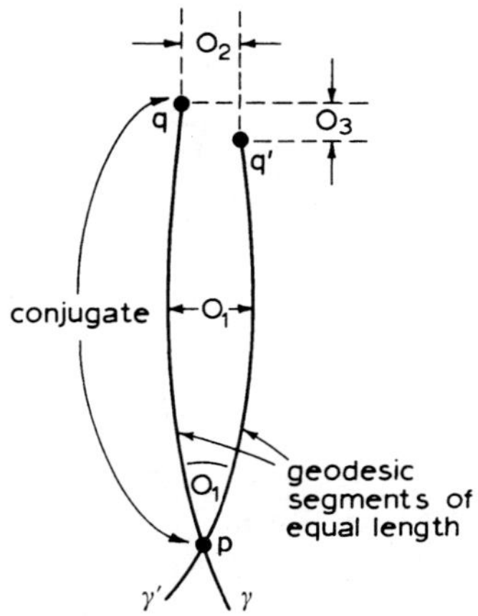

FIG. 47. *The points p and q are conjugate on γ; p is kept fixed and γ varied to a location γ' nearby, such that pq' and pq have equal length. The orders of separation are as indicated, with O_r standing for $O(\varepsilon^r)$*

for the reader who wishes to pursue this line of argument further. A rather different (and more satisfactory) way of approaching the problem will be given in 7.27. But before considering this, it will be useful to introduce the following slightly more general concept.

7.10. DEFINITION. Let γ be a timelike geodesic meeting, a smooth spacelike hypersurface Σ orthogonally at the point p. Then a point q is said to be *conjugate to Σ on γ* if and only if a nontrivial Jacobi field exists on γ which vanishes at p but not everywhere along γ, and which arises from a 1-parameter system of a. p. geodesics which are all orthogonal to Σ at their intersections with Σ (see Fig. 48).

7.11. *Remark.* If γ had been a spacelike geodesic, the situation would have been essentially the same, but with Σ a timelike hypersurface. However, for a *null* geodesic γ the situation is rather different. This is because if a null geodesic γ meets a *null hypersurface* Σ orthogonally at one point p, then Σ has to *contain* γ —or at least some finite portion of γ in the neighborhood of p, in case γ extends beyond the boundary ("edge") of Σ. (This is a familiar property of null hypersurfaces, which we shall return to in 7.13. We can choose Σ to be one of a family of null hypersurfaces, defined by u = const., with u a scalar field on M. We have

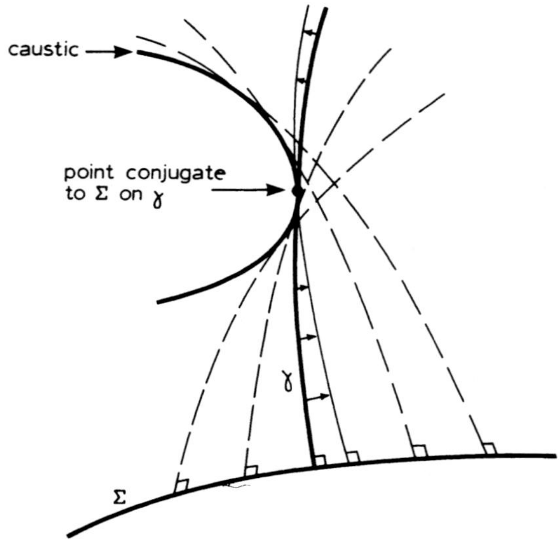

FIG. 48. *A point conjugate to a spacelike hypersurface Σ on a timelike geodesic γ orthogonal to Σ. The locus of such conjugate points as γ varies is called a caustic*

$T^aT_a = 0$, where the vector $T_a = \nabla_a u$ is null and normal to the hypersurfaces; therefore also tangent to the hypersurfaces: $T^a\nabla_a u = 0$. The T's are tangent to null geodesics lying on these hypersurfaces since $T^a\nabla_a T_b = T^a\nabla_a\nabla_b u = T^a\nabla_b\nabla_a u = T^a\nabla_b T_a = \frac{1}{2}\nabla_b(T^aT_a) = 0$. Exercise: write this simple calculation out using the index-free notation!) Thus it becomes somewhat confusing to talk about a point of γ being conjugate to the hypersurface Σ since the conjugate points would have to lie on Σ itself or on Σ extended. Actually the conjugate points would all be singular points of Σ (extended), constituting a cuspidal 2-surface (an $(n-2)$-surface if M is n-dimensional) called the *caustic* of Σ (see Fig. 49).

We can regain the usefulness of the conjugate point-to-surface concept for a null geodesic γ if instead of referring to a hypersurface Σ we use a spacelike 2-surface of "cross-section" of Σ. (This would be an $(n-2)$-surface if M were n-dimensional.) Null geodesics orthogonal to spacelike 2-surfaces constitute a situation which it is often useful to consider in relativity theory. The null geodesic generators of a null hypersurface Σ are, for example, orthogonal to *any* spacelike "cross-section" of Σ. Also, if an achronal set S happens to be a smooth spacelike hypersurface, it may well be that edge S is a smooth spacelike 2-surface, the union $S \cup$ edge $S = \bar{S}$ constituting a manifold with boundary embedded in M. (For example, let S be $t = 0$, $x^2 + y^2 + z^2 < 1$ in Minkowski space, edge S being $t = 0$, $x^2 + y^2 + z^2 = 1$.) Then, in the neighborhood of edge S, the hypersurface (with boundary) $\partial I^+[S] - S$ is smooth and null, being generated by null geodesics which meet edge S orthogonally (since edge S is locally a smooth spacelike "cross-section" of $\partial I^+[S]$). (In the above example, $\partial I^+[S] - S$ is $t \geqq 0$, $x^2 + y^2 + z^2 = (t+1)^2$, being generated by null lines $lx + my + nz - 1 = t \geqq 0, x:y:z = l:m:n$, where l, m, n are constant with $l^2 + m^2 + n^2 = 1$.) A related situation

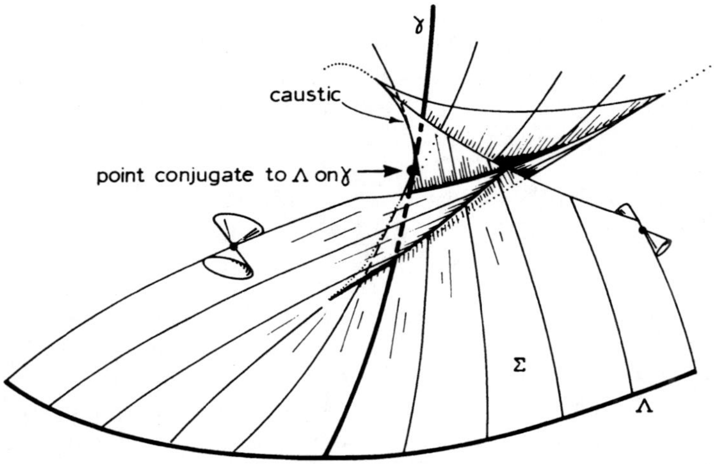

FIG. 49. *The null hypersurface Σ is generated by null geodesics one of which is γ; Λ is a spacelike 2-surface, being a cross-section of Σ, so γ is orthogonal both to Σ and to Λ. A point on γ conjugate to Λ lies on the caustic, at which Σ becomes locally singular*

It is worthwhile to examine $B = \partial I^+[\Lambda]$ in this example. Only the lowermost "spike" of the caustic can lie in B. The rest of the caustic lies in $I^+[\Lambda]$. The part of Σ which is contained in B consists of that up to and including the cross-over region, but not beyond it

is that arising from an achronal set S which happens itself to be a smooth spacelike 2-surface (e.g., $t = 0, x^2 + y^2 + z^2 = 1$ in Minkowski space). Then S (=edge S) is locally the intersection of two null hypersurfaces, these being, near S, portions of $\partial I^+[S]$. (In this second example these are $x^2 + y^2 + z^2 = (t + 1)^2, t \geqq 0$ and $x^2 + y^2 + z^2 = (t - 1)^2, 1 \geqq t \geqq 0$.) In fact, these two null hypersurfaces (extended) are simply the hypersurfaces traced out by the null geodesics which meet S orthogonally.

7.12. DEFINITION. Let γ be a null geodesic meeting a smooth spacelike 2-surface Λ orthogonally at the point p. Then a point $q \in \gamma$ is said to be *conjugate to* Λ *on* γ if and only if a nontrivial Jacobi field exists on γ which vanishes at p but not everywhere along γ, and which arises from a 1-parameter system of a. p. null geodesics which are all orthogonal to Λ at their intersections with Λ.

7.13. DEFINITION. There is an alternative way of thinking about conjugate points which is useful in some contexts. Suppose, that γ_0 is a timelike geodesic orthogonal to a spacelike hypersurface Σ. We consider the *congruence* of timelike geodesics (γ) which meet Σ orthogonally. We are concerned only with those members of the congruence which lie in some neighborhood Q of γ_0 in M. Then provided γ_0 does not extend too far away from Σ, we can choose Q small enough that the unit future-pointing tangent vectors to the geodesics (γ) constitute a smooth *vector field* T in Q. Since unit tangent vectors to geodesics are parallelly propagated along the geodesics we have $T^a\nabla_a T^b = 0$ (that is, $\nabla_T T = 0$); furthermore, the *rotation* of (γ) vanishes: $\nabla_a T_b - \nabla_b T_a = 0$ (i.e., $d(gT) = 0$). This last property follows from the fact that we can set $T_a = \nabla_a t \; (gT = dt)$, where the scalar field t measures the distance (i.e., "time") along γ from Σ, this being a consequence of the fact that the connecting vectors from points of γ to points of neighboring γ's (each parameterized by t) must be orthogonal to T all along γ, by 1.16, since they are orthogonal at Σ. Other quantities of interest concerning the congruence are the *divergence* $\theta = \nabla_a T^a$ ($\theta = \operatorname{div} T$) and the *shear* $\nabla_a T_b + \nabla_b T_a - \frac{2}{3}\theta(g_{ab} - T_a T_b)$, (i.e., $2 \operatorname{sym} \nabla T - \frac{2}{3}\theta(g - gT \otimes gT)$. We shall be concerned with the divergence particularly, later.

The parameter t can be used as a time coordinate in a *synchronous coordinate system* (cf. the proof in 7.2), the metric taking the form

$$ds^2 = dt^2 - q_{\alpha\beta}\, dx^\alpha\, dx^\beta, \qquad \alpha, \beta = 1, 2, 3,$$

where the components $q_{\alpha\beta}$ constitute a symmetric positive definite (3×3)-matrix (of functions of t, x^1, x^2, x^3). To set up this coordinate system, we let x^1, x^2, x^3 be arbitrary coordinates on Σ, which we label $t = 0$. Then the coordinates in the rest of Q are defined by taking x^α constant along each γ and t to measure distance from Σ along γ. Each hypersurface $t =$ const. will be orthogonal to each γ. The construction given in the proof of 7.2 is really a limiting case of this in which the hypersurface Σ degenerates into a point p, the congruence (γ) now consisting of timelike geodesics through p. The region in which the synchronous coordinate system is valid must now exclude the point p. The significance of all this, as we shall see in 7.26, is that the synchronous coordinate system is always valid, near γ_0, until a conjugate point to Σ, or p, is reached.

When γ_0 is a null geodesic, orthogonal to a spacelike 2-surface Δ_0, we must proceed somewhat differently. We can choose a null hypersurface Ω_0 to be generated by null geodesics (γ) orthogonal to Λ_0 (and belonging to the system continuous with γ_0), contained in some neighborhood Q of γ_0 in M. The fact that the hypersurface Ω_0 constructed in this way is *null* (provided γ_0 does not extend too far beyond Λ_0) follows from considerations similar to those described above. (All connecting vectors are orthogonal to the tangent vectors to γ, these being null, so the normal vectors to Ω_0 must be null.) We can construct a suitable *congruence* of null geodesics by allowing Λ_0 to vary smoothly in some 1-parameter family (Λ), parameter u, with $u = 0$ giving Λ_0. (We must move Λ in a direction not contained in Ω_0, i.e., not orthogonal to (γ)). This gives us a 1-parameter family of null hypersurfaces (Ω), and we extend the system of null geodesics (γ) to the congruence of null geodesics in Q which generate (Ω). We can choose our tangent vectors to (Ω) to constitute the *null vector field* T, given by $T_a = \nabla_a u$ ($gT = du$), regarding the parameter u as a scalar field on M, defined in Q, where $u =$ const. gives the hypersurfaces (Ω). The tangent vectors T are then parallelly propagated along (γ): $T^a \nabla_a T^b = 0$ ($\nabla_T T = 0$), cf. 7.11; and they are rotation-free: $\nabla_a T_b - \nabla_b T_a = 0$, ($d(gT) = 0$). The divergence of (γ) is given by $\theta = \nabla_a T^a = \operatorname{div} T$.

An analogue of a synchronous coordinate system gives the metric in Q in the form

$$ds^2 = 2\, du(dv + \tfrac{1}{2}a\, du + b_\lambda\, dx^\lambda) - r_{\lambda\mu}\, dx^\lambda\, dx^\mu, \qquad \lambda, \mu = 2, 3.$$

Here, v is chosen so that $v = 0$, $u =$ const. give the surfaces (Λ) (with $v = 0 = u$ giving Λ_0) and v is the affine parameter on γ corresponding to T. The remaining coordinates x^1, x^2 are chosen so that each geodesic γ is given by $u, x^1, x^2 =$ const.

This leads us to a metric in the above form, where the b_λ and $r_{\lambda\mu}$ are functions of u, v, x^1, x^2, the $r_{\lambda\mu}$ constituting a symmetric positive definite (2×2)-matrix. A coordinate system of the above type will be called a *null* coordinate system. A particular case is obtained if we allow the surfaces (Λ) to degenerate into points. Again, as we shall see in 7.26, the significance of all this is that the coordinate system remains valid in the neighborhood of γ_0 until a conjugate point to Λ_0 is reached.

7.14. PROPOSITION. *With the notation of* 7.13, *the divergence* $\theta = \nabla_a T^a$ *satisfies* $D\Delta = \theta\Delta$, *where* $D = T^a\nabla_a$ $(= \nabla_T)$ *and where* Δ *is proportional to an element of volume on* Σ *if* γ *is timelike, or an element of surface area on* Λ *if* γ *is null,* Δ *being traced out by the geodesics* (γ) (i.e., *Lie propagated along* γ) *as* Σ *or* Λ *varies.*

Proof. If γ is timelike, set up a synchronous coordinate system as in 7.13, and put $X_0 = T = \partial/\partial t$, $X_\alpha = \partial/\partial x^\alpha$, $\alpha = 1, 2, 3$. If γ is null use a null coordinate system as in 7.13 and put $X_0 = T = \partial/\partial v$, $X_1 = \partial/\partial u$, $X_\lambda = \partial/\partial x^\lambda$, $\lambda = 2, 3$. The 4-volume spanned by the coordinate vectors $X_0, \cdots, X_3$ is given by $(|\det g_{\lambda\tau}|)^{1/2} = \Delta_4$, where $g_{\lambda\tau}$ are the components of the metric tensor. In the two cases this matrix is

$$\begin{pmatrix} 1 & 0 & 0 & 0 \\ 0 & & & \\ 0 & & -q_{\alpha\beta} & \\ 0 & & & \end{pmatrix} \quad \text{or} \quad \begin{pmatrix} 0 & 1 & 0 & 0 \\ 1 & a & b_2 & b_3 \\ 0 & b_2 & & \\ 0 & b_3 & & -r_{\lambda\mu} \end{pmatrix},$$

respectively, so setting $\Delta = (-\det q_{\alpha\beta})^{1/2}$ or $\Delta = (-\det r_{\lambda\mu})^{1/2}$, in accordance with the conditions of the proposition, we clearly have $\Delta_4 = \Delta$ in each case. Finally, the formula $D\Delta_4 = \theta\Delta_4$ may be seen in various ways, for example, by calling upon the well-known classical formula

$$\left\{ \begin{matrix} \rho \\ \rho\sigma \end{matrix} \right\} = \tfrac{1}{2}(\partial/\partial x^\sigma) \log|\det g_{\kappa\tau}| .$$

7.15. *Remark.* It should be observed that the concept of the T field, in the neighborhood of the causal geodesic γ_0 can still make sense even beyond a region of breakdown of the synchronous or null coordinate system. All the vectors $X_0, \cdots, X_3$ introduced in the proof in 7.14 can be defined everywhere along γ_0, even beyond conjugate points, since the X_σ are just particular Jacobi fields. Thus we can define Δ all along γ (since the $g_{\kappa\tau}$ are just the scalar products: $g_{\kappa\tau} = g(X_\kappa, X_\tau)$). We have to be careful about differentiating further, however, to obtain θ, say, as the next result shows.

7.16. PROPOSITION. *With the notation of* 7.13 *and* 7.14, *a point* w *is conjugate to* Σ *or* Λ_0 *on* γ_0 *if and only if* $\Delta = 0$ *at* w. *Furthermore,* θ *exists and is continuous at all points of* γ_0 *at which* $\Delta \neq 0$, *while* θ *becomes unbounded near any point* w *at which* $\Delta = 0$, *with* θ *large and positive just to the future of* w, *and large and negative just to the past of* w *on* γ_0.

Proof. A conjugate point w to Σ_0 or Λ_0 is given when a nontrivial Jacobi field Y on γ_0 vanishes at w and connects γ_0 to a "neighboring geodesic" γ also orthogonal to Σ_0 or Λ_0. Clearly such a Y must be a nontrivial linear combination of $X_0, \cdots, X_3$. Thus, the vectors $X_0, \cdots, X_3$ must be linearly dependent at the point w, and this fact actually characterizes w as a conjugate point. The linear dependence can be expressed as $\Delta = 0$, so the first part of 7.16 is established. The second part follows from the relation $D \log \Delta = \theta$ of 7.14.

7.17. *Remark.* It is clear that we can allow Σ_0 or Λ_0 to degenerate to a point, p, in 7.16 and the result remains true. Note that at p itself each of X_2 and X_3 vanishes and, if γ is timelike, X_1 also vanishes at p. Wherever $X_\lambda = 0$ we must have $DX_\lambda \neq 0$ (with $D = \nabla_T = T^a\nabla_a$), because otherwise we should have $X_\lambda = 0$ along γ_0 (cf. 1.15). It follows that $\Delta \sim t^3$ at p if γ_0 is timelike and $\Delta \sim v^2$ at p if γ_0 is null.

7.18. *Remark.* By the same token we have $\Delta \sim (t - t_1)^r$ or $(v - v_1)^r$ at *any* point at which r linearly independent combinations of the X_λ vanish (i.e., at which there are 4-r linearly independent X_ρ's). Such a point is said to have *conjugate degree r* (with respect to Σ_0, Λ_0, or p). Notice also that conjugate points (to Σ_0, Λ_0 or p) on γ_0 must be *isolated*, that is, they cannot accumulate at any point (conjugate or otherwise) of γ_0. At worst, we can have r conjugate points *coincident* at one point of γ_0 ($r \leqq 3$ if γ is timelike, $r \leqq 2$ if γ is null). This is the intuitive meaning of the conjugate degree.

7.19. PROPOSITION. *If γ_0 is timelike, then, with notation as in* 7.13, 7.14, *taking $\Delta > 0$ for simplicity, we have*

$$D^2\Delta^{1/3} \leqq \tfrac{1}{3}R_{ab}T^aT^b\Delta^{1/3},$$

while if γ_0 is null, we have

$$D^2\Delta^{1/2} \leqq \tfrac{1}{2}R_{ab}T^aT^b\Delta^{1/2}.$$

(The Ricci tensor is defined by $R_{ab} = R^c{}_{acb}$

Proof. Ricci identities applied to $T^a\nabla_a\nabla_bT^b - T^a\nabla_b\nabla_aT^b$ give Raychaudhuri's equation [31], [32], [9]:

$$\begin{aligned} D\theta &= R_{ab}T^aT^b - \nabla_aT^b\nabla_bT^a \\ &= R_{ab}T^aT^b - (\nabla_aT_b)(\nabla^aT^b) \end{aligned}$$

(using $T^a\nabla_aT^b = 0, \nabla_bT^b = \theta, \nabla_aT_b = \nabla_bT_a$; cf. 7.13). Suppose T is timelike. Then express Raychaudhuri's equation thus:

$$D\theta - \tfrac{1}{3}\theta^2 = R_{ab}T^aT^b - S_{ab}S^{ab},$$

where

$$S_{ab} = S_{ba} = \nabla_aT_b - \tfrac{1}{3}\theta(g_{ab} - T_aT_b),$$

is the shear tensor. From the fact ($S_{ab}T^a = 0, S_{ab}T^b = 0$) that S_{ab} has all its components in the (negative definite) spacelike hyperplane orthogonal to T, we obtain $S_{ab}S^{ab} \geqq 0$. Furthermore, by 7.14, $D^2\Delta^{1/3} = D(\frac{1}{3}\theta\Delta^{1/3}) = \frac{1}{3}\Delta^{1/3}(D\theta - \frac{1}{3}\theta^2)$ so the result for timelike γ_0 follows. Now suppose T is null and rewrite Raychaudhuri's equation:

$$D\theta - \tfrac{1}{2}\theta^2 = R_{ab}T^aT^b - \sigma_{ab}\sigma^{ab},$$

where

$$\sigma_{ab} = \sigma_{ba} = \nabla_a T_b - \tfrac{1}{2}\theta\gamma_{ab},$$

with γ_{ab} $(= \gamma_{ba})$ defining the negative definite intrinsic 2-metric of Λ_0, that is,

$$g_{ab} = T_aN_b + N_aT_b + \gamma_{ab},$$

where T and N are null vectors orthogonal to γ_0 normalized so that $T^aN_a = 1$. We have $\gamma_{ab}\gamma^{ab} = 2$, $\gamma^{ab}\nabla_a T_b = \theta$. We have

$$\begin{aligned}\sigma_{ab}\sigma^{ab} &= \sigma^{cd}\sigma^{ab}g_{ca}g_{db}\\ &= \sigma^{cd}\sigma^{ab}\gamma_{ca}\gamma_{db} \geqq 0,\end{aligned}$$

by the negative definiteness of the 2-metric, so the required result for the null case follows similarly to the timelike case above.

7.20. *Remark.* In an n-dimensional space-time the result of 7.19 would be still valid, with $1/(n-1)$ replacing $\frac{1}{3}$ and $1/(n-2)$ replacing $\frac{1}{2}$. The proof is essentially unaffected.

7.21. *Remark.* In a "physically reasonable" space-time subject to Einstein's equations it is normally supposed that $R_{ab}T^aT^b \leqq 0$, since this inequality represents a very reasonable restriction on the energy-momentum density of the matter. This is called the *energy condition* (or the *strong energy condition* if the inequality is required to hold for all timelike T and called the *weak energy condition* if the inequality is required merely for all null T). Then 7.19 can be strengthened to $D^2\Delta^{1/3} \leqq 0$ along timelike geodesics and $D^2\Delta^{1/2} \leqq 0$ along null geodesics, provided $\Delta > 0$. This has the effect, of prime importance for the singularity theorems, that once the geodesics of the congruence (γ) start to converge, then they must, within a finite value of the affine parameter, inevitably converge to a caustic ($\Delta = 0$)—assuming that γ_0 is a complete geodesic (see Fig. 50). This is called the Raychaudhuri [31] (or Raychaudhuri–Komar [32]) effect within the

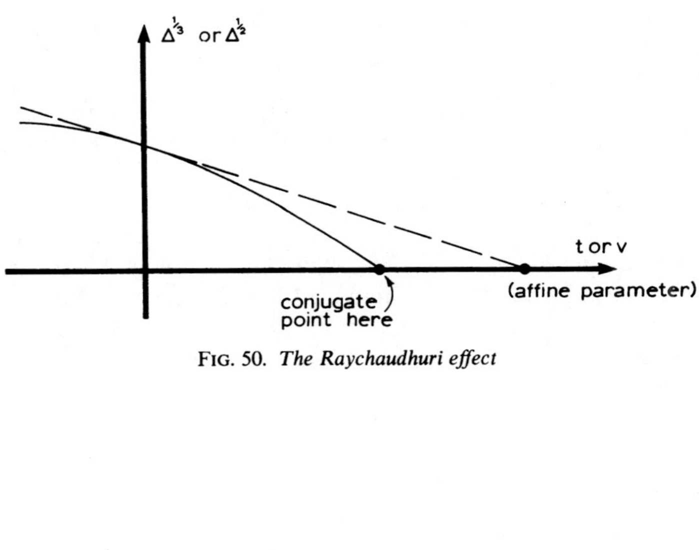

FIG. 50. *The Raychaudhuri effect*

context of relativity theory. For manifolds with a positive definite metric, essentially the same effect had been studied earlier by Myers [33]. Even in the absence of an energy condition an effect of this type persists, as the next proposition shows.

7.22. PROPOSITION. *Let* $\alpha = \frac{1}{3}$ *if* γ_0 *is timelike and* $\alpha = \frac{1}{2}$ *if* γ_0 *is null* (*with notation as in* 7.13, 7.14) *and let* $\Delta^\alpha = A$, $D\Delta^\alpha = -B$, *at some point* a *on* γ_0. *Suppose that* $\alpha R_{cd}T^cT^d \leqq k^2$ *throughout the segment* ab *of* γ_0, *where the parameter value* (t *or* v) *at* b, *on* γ_0, *is greater than that at* a *by at least the amount* $k^{-1}\tanh^{-1}(Ak/B)$. *Then* $\Delta = 0$ *somewhere on* ab.

Proof. We compare the equation $D^2\Delta^\alpha \leqq k^2\Delta^\alpha$ with the explicit solution of the equation $D^2x = k^2x$ for which $x = A$, $Dx = -B$ at a. Then the result is straightforward.

7.23. *Remark.* The result 7.22 shows that, on a complete geodesic γ_0, if we can be sure that $B > Ak$ (taking $A, k \geqq 0$), then γ_0 encounters a caustic somewhere to the future of a.

7.24. PROPOSITION. *Let* γ *be a causal geodesic from* p *to* q. *Suppose that either*

(a) q *is conjugate to* p *on* γ,

or

(b) γ *is orthogonal at* p *to a smooth spacelike hypersurface* Σ (γ *timelike*) *or 2-surface* (γ *null*) *and* q *is conjugate to* Σ *on* γ.

Then there is a first (*i.e., pastmost*) *point* q', *to the future of* p *on* γ, *with property* (a) *or* (b), *respectively, and which varies continuously with* p *and* γ (Σ *being kept fixed in case* (b), *for simplicity*).

Proof. The existence of a first conjugate point q' is a consequence of the fact (cf. 7.18) that conjugate points are isolated. Now we saw, in 1.15, that a Jacobi field on γ is a solution of the equation $D^2V^a = R^a{}_{bcd}T^bV^cT^d$. The solutions of this equation are continuous functions of the initial data for V, namely of the values of V and DV at any one point of γ. Furthermore, if we allow $R^a{}_{bcd}$ to vary, then the solutions will vary continuously as functions of $R^a{}_{bcd}$ also. Allowing γ to vary has the same effect as this. What we have to show is that this implies that conjugate points vary continuously also.

Let r be a point of $\gamma = \gamma_0$ which lies on the caustic of a congruence (γ) containing γ_0. With the notation of 7.13, 7.14 we have $\Delta = 0$ at r. By 7.18, $\Delta^{1/3} \sim (t - t_1)^{1/3}$ or $(t - t_1)^{2/3}$ or $t - t_1$, if γ_0 is timelike, and $\Delta^{1/2} \sim (v - v_1)^{1/2}$ or $v - v_1$, if γ is null. In every case the power is not greater than unity. Thus we can invoke 7.22 to show that for any sufficiently small interval of γ_0 about r we can choose points a and b in the interval and ensure that $\tilde{\Delta} = 0$ somewhere between a and b, for any congruence ($\tilde{\gamma}$) which differs sufficiently little from (γ) and for any $\tilde{R}^a{}_{bcd}$ which differs sufficiently little from $R^a{}_{bcd}$. To supply all the details for this argument would be rather tedious (exercise).

7.25. *Remark.* The essential feature of 7.24 is the fact that conjugate points on a causal geodesic cannot *annihilate* one another as the circumstances vary continuously. For this reason, 7.22 was invoked. The quantity Δ cannot approach zero too closely, so to speak, without actually becoming zero. It is curious that the argument as given depends on an inequality resulting from the negative definiteness of the appropriate orthogonal subspace. It would be interesting to know whether the result for spacelike geodesics is even true. That is, can conjugate points on a spacelike geodesic annihilate one another?

7.26. PROPOSITION. *With notation as in* 7.13, *if* γ_0 *contains no conjugate point* (*to* Σ_0, Λ_0, *or* p), *then there is a synchronous coordinate system if* γ_0 *is timelike, or a null coordinate system if* γ_0 *is null, which is valid in some neighborhood of* γ_0, *or, in the case of geodesics through* p, *in some neighborhood of the portion of* γ_0 *to the future of* p.

Proof. By 7.24 we are assured of the existence of a neighborhood Q of γ_0 which does not intersect the caustic of (γ) (except at p, in the case of geodesics through p) and indeed throughout which the congruence is actually defined (and one-valued). It should be clear that the construction given in 7.13 then actually yields coordinate systems of the required type. (Exercise: supply the details!)

7.27. THEOREM [4], [9]. *Let* γ *be a causal geodesic from* p *to* q.

(a) *If* γ *contains an internal point which is conjugate to* p (*or to* q), *then there is a causal trip from* p *to* q *of length strictly greater than that of* pq (*so if* γ *is null, then* $p \ll q$).

(b) *Let* Σ *be a hypersurface if* γ *is timelike, or a 2-surface if* γ *is null, which is spacelike and contains* p, *such that either* γ *is not orthogonal to* Σ *at* p, *or else it is orthogonal and there is a conjugate point to* Σ *between* p *and* q *on* γ. *Then there is a causal trip from a point of* Σ *to* q, *of length strictly greater than that of* pq (*so if* γ *is null, then* $q \in I^+[\Sigma]$).

Proof. Let r be the first conjugate point to Σ, or to p, beyond Σ. Suppose, first, that γ is timelike and (in the case (b)) orthogonal to Σ. Then by 7.26 we can set up a synchronous coordinate system $\mathfrak{S}$ valid in some neighborhood of the portion of γ between p and r; and valid also at p, in case (b). In each case the t coordinate measures distance from p (case (a)) or Σ (case (b)) along timelike geodesics through p (case (a)) or orthogonal to Σ (case (b)). Choose a point w on γ, to the future of r, which is close enough to r that the segment rw contains no pair of conjugate points. Then if r' precedes r on γ and is close enough to r, the point r' will not be conjugate to w either. Thus, the segment $r'w$ is covered (except for w) by another synchronous coordinate system $\hat{\mathfrak{S}}$ whose $\hat{t}$ coordinate measures minus the distance to w. The vectors $\hat{T}^a = \nabla^a \hat{t}$ are future-pointing unit timelike vectors, with $\hat{T} = T$ along $\gamma = \gamma_0$. ($T^a = \nabla^a t$; i.e., $T = g^{-1}\, dt = \partial/\partial t$.) Now since r is conjugate to p or to Σ, there is a nontrivial Jacobi field X on $\gamma = \gamma_0$ which vanishes at r and arises from a 1-parameter subfamily, containing γ_0, of the congruence (γ) of time-lines of $\mathfrak{S}$. We have $DX \neq 0 = X$ at r. So X has the form $X = (t_0 - t)Y$, where t_0 is the t-value at r (i.e., the distance $l(pr)$). Y is a smooth vector field defined along γ_0 which is orthogonal to γ_0 and which is nonvanishing at r, so Y is spacelike at r: $Y^a Y_a < 0$. Now

$$\begin{aligned} Y^a Y^b \nabla_a \nabla_b t &= Y^a Y^b \nabla_a T_b \\ &= (t_0 - t)^{-1} Y^b X^a \nabla_a T_b \\ &= (t_0 - t)^{-1} Y^b D X_b = (t_0 - t)^{-1} Y^b D\{(t_0 - t) Y_b\} \\ &= -(t_0 - t)^{-1} Y^b Y_b + Y^b D Y_b \end{aligned}$$

near r, this being large and positive just to the past of r on γ_0. Thus, at a point r' sufficiently close to r, just before r on γ_0 we shall have

$$Y^a Y^b \nabla_a \nabla_b (t - \hat{t}) > 0$$

since $\nabla_a \nabla_b \hat{t}$ is well-behaved at r'. Also,

$$Y^a \nabla_a (t - \hat{t}) = 0$$

at r (since $T = \hat{T}$ at r). Now consider the a. p. geodesic η with tangent vector Y at r'. We have $Y^a \nabla_a Y^b = 0$ along η, so the mth derivative with respect to the parameter on η, of a function φ defined on η can be re-expressed as

$$(Y)^m(\varphi) = (Y^a \nabla_a)^m \varphi = Y^{a_1} Y^{a_2} \cdots Y^{a_m} \nabla_{a_1} \nabla_{a_2} \cdots \nabla_{a_m} \varphi.$$

Hence, by the two previous formulae above (and by Taylor's theorem), it follows, setting $\varphi = t - \hat{t}$, that at any point r'' near enough to r' on η (and so covered by both $\mathfrak{S}$ and $\hat{\mathfrak{S}}$) the value of $t - \hat{t}$ must exceed the value of $t - \hat{t}$ at r'. But $t - \hat{t} = t + (-\hat{t})$, at r'', is the total length of the trip v consisting of the relevant member of (γ) up to r'', together with the geodesic $r''w$ (Fig. 51). The length of v thus exceeds

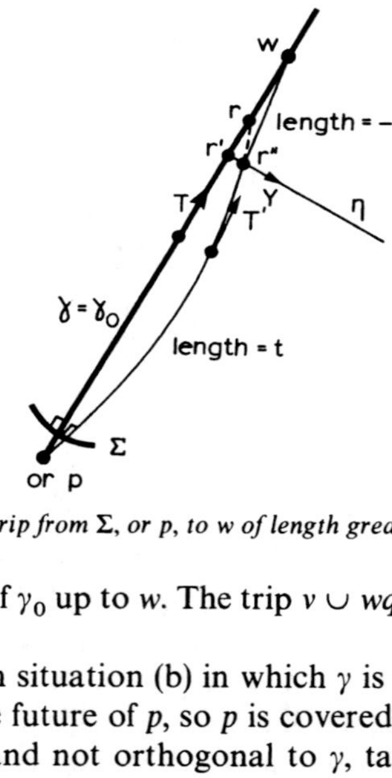

FIG. 51. *How to construct a trip from Σ, or p, to w of length greater than that of γ to w (for* 7.27)

the length of the portion of γ_0 up to w. The trip $v \cup wq$ therefore has length greater than $l(\gamma_0)$ as required.

Next, suppose we are in situation (b) in which γ is timelike but not orthogonal to Σ. Choose w just to the future of p, so p is covered by $\hat{\mathfrak{S}}$. Choose a vector Z at p which is tangent to Σ and not orthogonal to γ, taken in the direction so that $0 > Z^a T_a = Z^a \nabla_a \hat{t}$ at p. Then, if we choose p', near enough to p, on some curve on Σ with tangent vector Z at p, we shall have $-\hat{t}$ at p' greater than $-\hat{t}$ at p (since $Z(-\hat{t}) = Z^a \nabla_a(-\hat{t}) > 0$ at p). Since $-\hat{t}$ measures distance to w, we have $l(p'w \cup wq) > l(\gamma)$ as required.

When γ is null (and orthogonal to Σ, in case (b)), the argument, to begin with, follows closely the one given above except that null coordinate systems $\mathfrak{S}$ and $\hat{\mathfrak{S}}$ replace the synchronous ones used before, where we scale the u, $\hat{u}$ coordinates so that $\nabla_a u = T_a = \hat{T}_a = \nabla_a \hat{u}$ along $\gamma = \gamma_0$, and take $u = \hat{u} = 0$ along γ_0. (We have $T = g^{-1}\,du = \partial/\partial v$.) In place of $t_0 - t$ we have $v_0 - v$, but in place of t we have u; that is, we take $X = (v_0 - v)Y$, with v_0 the affine parameter value of r, and our calculation becomes

$$Y^a Y^b \nabla_a \nabla_b u = \cdots = -(v_0 - v)^{-1} Y^b Y_b + Y^b D Y_b.$$

The vector Y is still spacelike, being orthogonal to T (and not proportional to T since such a Jacobi field would have to be constant). As before, we choose r' preceding r on γ_0, and sufficiently close to r that

$$Y^a Y^b \nabla_a \nabla_b u > Y^a Y^b \nabla_a \nabla_b \hat{u}$$

at r'. Now set $U = \partial/\partial u$ at r' and consider $\exp_{r'}$ applied to the plane π spanned by U and Y. Write the general element of π as $xU + yY$ and consider each of u and $\hat{u}$ as functions of the coordinates (x, y). We have $u = \hat{u} = 0$ at the origin $r'(0, 0)$, and also $\partial u/\partial y = \partial \hat{u}/\partial y = 0$ at r' (since $Y^a \nabla_a u = Y^a T_a = 0$ and similarly for $\hat{u}$). Also we have $\partial u/\partial x = \partial \hat{u}/\partial x = 1$ at r' (since $U(u) = \partial u/\partial u = 1$ and $U(\hat{u}) = U^a \nabla_a \hat{u} = U^a \hat{T}_a = U^a T_a = U(u) = 1$). Finally, the above displayed formula states $A > \hat{A}$, where $A = [\partial^2 u/\partial y^2]_{r'}$, and $\hat{A} = [\partial^2 \hat{u}/\partial y^2]_{r'}$. Consider the curve $4x + (A + \hat{A})y^2 = 0$: Taylor's theorem gives[3] $u = \frac{1}{4}(A - \hat{A})y^2 + o(y^2)$, $\hat{u} = \frac{1}{4}(\hat{A} - A)y^2 + o(y^2)$, so for small enough $y > 0$ we have $\hat{u} < 0 < u$. Thus $\exp_{r'}$ applied to this value of (x, y) gives us a point r'' for which $\hat{u} < 0$, so $r'' \ll w$, and for which $u > 0$, so $p \ll r''$ in case (a) and $r'' \in I^+[\Sigma]$ in case (b). We have $w \prec q$, whence $p \ll q$ in case (a) and $q \in I^+[\Sigma]$ in case (b), as required.

Finally, in case (b), if γ is not orthogonal to Σ, the argument is similar to that for the timelike case, except that $\hat{\mathfrak{S}}$ is null and not synchronous. We obtain $-\hat{u} > 0$ for a point p' on Σ, so $p' \ll w$. Thus $w \in I^+[\Sigma]$, so $q \in I^+[\Sigma]$ as required.

7.28. *Remark.* The behavior of geodesics of fixed length near a conjugate point in a positive definite space M and hyperbolic normal space-time M (γ timelike) are compared in Fig. 52. Some of the essential complication of the situation is depicted. This indicates something of the difficulties which stand in the way of successfully completing the intuitive arguments of 7.9 (cf. Figs. 46 and 47).

As a converse to the theorem in 7.27 we have the following proposition.

7.29. PROPOSITION. *Let γ be a causal geodesic from p to q such that either*:

(a) *γ contains no pair of conjugate points*;

or

(b) *Σ is a hypersurface if γ is timelike, or a 2-surface if γ is null, which is spacelike and is orthogonal to γ at p, and is such that no point of γ is conjugate to Σ.*

Then there is a neighborhood Q of γ, in M, such that any causal curve η, in Q, from p to q (case (a)) *or from a point of Σ to q (case* b)) *satisfies $l(\eta) \leqq l(\gamma)$, with equality holding only if $\eta = \gamma$.*

[3] This assumes C^2 differentiability of u, $\hat{u}$ only; similarly in the timelike case, we used just C^2 differentiability of t, $\hat{t}$. These conditions will hold (provided the metric g of M is C^2) if the (hyper-)surface Σ is C^2.

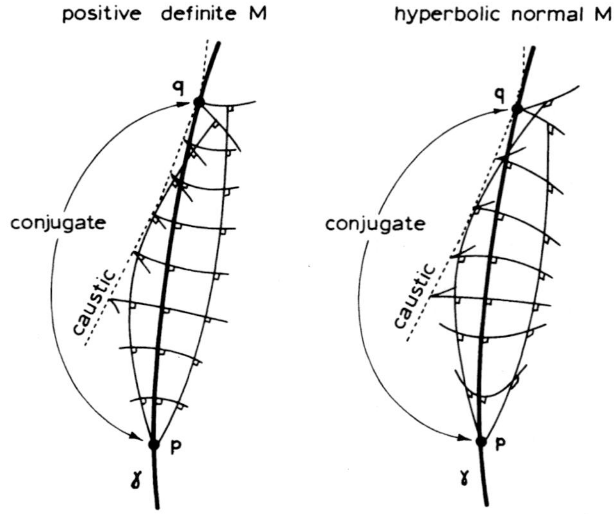

FIG. 52. *The points p and q are conjugate on a geodesic γ. As γ varies through p, points of equal distance from p along γ trace out curves orthogonal to γ. The situations for a positive definite space M and for a space-time M (γ timelike) are compared*

Proof. The existence of a suitable synchronous coordinate system (γ timelike) or null coordinate system (γ null), valid for some Q, is assured by 7.26. The proof proceeds as in 7.2 in the timelike case and similarly in the null case.

7.30. *Remark.* It is easy to construct examples of space-times containing causal curves with no conjugate points but which are not *globally* of maximum length (cf. the 2-dimensional Einstein cylinder in Fig. 14).

The cases not covered either by 7.27, or by 7.29, occur when q itself is conjugate to p or Σ. Here the situation can become complicated. "Generically," there will be causal geodesics of length greater than γ, from p, or Σ, to q in *any* neighborhood of γ. (One "follows back" along the caustic through q for a short while.) But in particular cases γ may still be maximal, either uniquely so, or sharing its maximality with other geodesics.

SECTION 8

Singularity Theorems

8.1. *Remark.* I shall only very briefly indicate some of the applications of the preceding theory here. For details the reader is referred back to the published literature. I shall state two theorems only and briefly indicate their method of proof and a few relevant lemmas.

8.2. THEOREM [4]. *If a space-time M satisfies the following two conditions, then there is a past-endless geodesic in M which has a finite length*:

(a) *M contains a closed*[1] *spacelike hypersurface* Σ, *the normals to which diverge at every point of* Σ (*i.e., the congruence of geodesics meeting* Σ *orthogonally have* $\theta > 0$ *at every point of* Σ),

(b) *the energy condition* (cf. 7.21) *holds at every point of M.*

Discussion of proof. Note, first, that if we had assumed that Σ was a Cauchy hypersurface for M (as had indeed been the assumption in an earlier version of the theorem [2]), then the proof would be very direct from the preceding work. For, by the Raychaudhuri effect (see 7.21), every past-endless geodesic γ orthogonal to Σ must, if it has infinite length into the past, contain a futuremost point q conjugate to Σ. Since Σ is compact and such conjugate points move continuously (7.24), there must be an upper bound B to the distance, along γ's, from q to Σ. But if every past-endless γ extends to indefinitely great distance, along γ, to the past of Σ, we must be able to find w on γ whose distance to Σ exceeds B. Thus q lies between w and Σ on γ, so by 7.27, γ is not maximal from w to Σ. This applies whichever γ is chosen through w. But with Σ a Cauchy hypersurface, we have $w \in D^-(\Sigma)$, so by 6.6 and 7.7 a causal geodesic of maximal length *must* exist from q to Σ, which leads to a contradiction.

If Σ is not taken to be a Cauchy hypersurface, we must study its Cauchy horizon. (It was for this purpose that Hawking introduced the Cauchy horizon concept.) But first it is necessary to show that Σ may be taken to be an achronal set. To this end, we need a lemma.

8.3. LEMMA. *If M contains a spacelike hypersurface* Σ *without boundary, then there is a covering manifold* M^*, *of M, with the property that* M^* *contains a discrete set of isomorphic copies of* Σ, *each of which is achronal in* M^*.

Proof. There are various different methods of achieving this (see [4], [11], [24], [9], or do as exercise).

Continuation of argument for 8.2. If Σ is not achronal in 8.2 we apply the argument instead to M^*. So we may suppose without loss of generality that Σ *is* achronal in M. The above argument shows that *if* each γ has infinite length into the past, then

[1] "Closed" means, here, that the 3-manifold Σ is "compact without boundary" (cf. "closed curve").

each γ must have points in $I^-[\Sigma] - D^-(\Sigma)$, so γ meets $H = H^-(\Sigma)$. In fact $D^-(\Sigma)$ must lie within the compact region swept out by portions of γ curves of length B, with future endpoints on Σ. Since Σ is closed, so is H (5.5). Therefore H is compact. Since Σ is spacelike and edgeless, $H \cap \Sigma = \varnothing$, so H is a compact C^0-manifold without boundary (3.17). Set $f = \gamma \cap H$ and define $p(f)$ to be the maximum of lengths of segments of γ from f to Σ (attained, for fixed f, owing to the compactness of Σ and (γ)). One can show that $p(f)$ attains its minimum for $f \in H$, say at $f = f_0$. Now, by 5.12, a null geodesic η on H has f_0 as past endpoint. Choose f_1 just to the future of f_0 on η. The length of the causal trip from f_0 to Σ, consisting of f_0f_1 and the maximal γ curve from f_1 to Σ cannot be less than $p(f_0)$. Take f_2 just to the future of f_1 on γ. Then $l(f_0f_2) > l(f_0f_1) + l(f_1f_2)$ (cf. 7.2; we can take f_1, f_2, f_3 all in one simple region) so we obtain a trip ζ from f_0 to Σ of length k, greater than $p(f_0)$. Take b on ζ close enough to f_0 that the distance from b to Σ exceeds $p(f_0)$. Let b approach f_0 and obtain a limiting γ through b (by compactness of Σ). The implied relation $p(f_0) \geqq k > p(f_0)$ is the required contradiction establishing 8.2.

8.4. *Remark.* The implication of 8.2 is that a spatially closed universe which is everywhere expanding ($\theta > 0$) must, if it satisfies the energy condition, possess an initial "singularity." This is recognized by the presence of incomplete past-endless timelike geodesics.

The final theorem uses the following lemmas.

8.5. LEMMA. *If γ is a null geodesic lying on $I^+[S]$ or on $H^+[S]$, for some achronal set $S \subset M$, then γ cannot contain a pair of conjugate points except possibly at its endpoints.*

Proof. Immediate from 7.27 and the achronality of $I^+[S]$ and $H^+[S]$.

8.6. LEMMA. *If M contains no closed trips and if every endless null geodesic in M contains a pair of conjugate points, then strong causality holds everywhere.*

Proof. The result follows from 4.31 (since case (e) holds), and 7.27 [5], [6].

8.7. DEFINITION. A *future-trapped set* is a nonempty achronal closed set $S \subset M$ for which $E^+(S) = J^+[S] - I^+[S]$ is compact. Any future-trapped set S must itself be compact, since $S \subset E^+(S)$. Observe that any closed spacelike hypersurface which is an achronal set, must be future-trapped, but this is a very special case. The time-reverse of a future trapped set is called *past-trapped*.

8.8. LEMMA [6]. *If S is a future-trapped set for which strong causality holds at every point of $I^+[S]$, then there exists a future-endless timelike curve (or trip) $\gamma \subset \operatorname{int} D^+(E^+(S))$.*

Outline of proof. See [6]. The argument is to show first that $H = H^+(E^+(S))$ is noncompact or empty. This is done by trying to cover H with a finite number of local causality neighborhoods and deriving a contradiction. Then a smooth timelike vector field on M is chosen (cf. 1.4) the integral curves of which establish a homeomorphism between $E^+(S)$ and H *unless* the required future-endless curve γ exists. The homeomorphism cannot exist since $E^+(S)$ is compact and nonempty. (Exercise: supply the details—or look them up!)

8.9. THEOREM [6]. *No space-time can satisfy the following three requirements together:*

(a) *M contains no closed trips,*

(b) *every endless causal geodesic in M contains a pair of conjugate points,*
(c) *there exists a future-trapped set* $S \subset M$.

Outline of proof. The idea is to employ the lemma in 8.8 to obtain a future-endless trip $\gamma \subset \operatorname{int} D^+(E^+(S))$ (using 8.6). Then set $T = \overline{I^-[\gamma]} \cap E^+(S)$ and show that T is past-trapped. The time-reverse of 8.8 then gives a past-endless trip $\alpha \subset \operatorname{int} D^-(E^-(T))$. We choose $a_0, a_1, a_2, \cdots$ receding into the past indefinitely and $c_0, c_1, c_2, \cdots \in \gamma$ proceeding into the future indefinitely, where $a_0 \ll c_0$. We obtain $J^+(a_i) \cap J^-(c_i)$ as compact sets throughout which strong causality holds, so 7.7 can be used to obtain a maximal causal geodesic μ_i from a_i to c_i. Taking limits appropriately we obtain a contradiction between 7.27 and condition (b). The details of this argument are left as an exercise for the reader (or look them up [6]).

8.10. *Remark.* The physical significance of condition (b) in 8.9 is that this condition is a consequence of the energy condition (cf. 7.21) if we impose, in addition to completeness for causal geodesics, a physically reasonable condition of "generality" on the space-time, this condition being one which any small amount of randomly oriented curvature (e.g., weak gravitational waves or static fields) would be sufficient to ensure. The significance of condition (c) is that such a set S would be expected to arise in suitable situations of gravitational collapse. Thus, the physical implication of the theorem is that "singularities" (i.e., causal geodesic incompleteness) would be expected to arise whenever such a collapse takes place. It is not my purpose, however, to enter into the physical ramifications of these results here. The interested reader is referred to the literature [4], [6], [9].

References

[1] S. W. HAWKING, Phys. Rev. Lett., 15 (1965), 689.
[2] ———, Proc. Roy. Soc. Lond., A294 (1966), 511.
[3] ———, Ibid., A295 (1966), 490.
[4] ———, Ibid., A300 (1967), 187.
[5] ———, *Singularities and the Geometry of Space-time*, Adams Prize Essay, Cambridge University, 1966.
[6] S. W. HAWKING AND R. PENROSE, Proc. Roy. Soc. Lond., A314 (1970), 529.
[7] R. PENROSE, Phys. Rev. Lett., 14 (1965), 57.
[8] ———, *An Analysis of the Structure of Space-time*, Adams Prize Essay, Cambridge University, 1966.
[9] ———, *Structure of Space-time*, Battelle Rencontres, 1967 Lectures in Mathematics and Physics C. M. DeWitt and J. A. Wheeler, eds., W. A. Benjamin, New York, 1968.
[10] R. GEROCH, Phys. Rev. Lett., 17 (1966), 446.
[11] ———, J. Math. Phys., 9 (1968), 1739.
[12] N. STEENROD, *The Topology of Fibre Bundles*, Princeton University Press, Princeton, N.J., 1951.
[13] L. MARKUS, Ann. Math., 62 (1955), 411.
[14] N. J. HICKS, *Notes on Differential Geometry*, D. van Nostrand, Princeton, N.J., 1965.
[15] C. W. MISNER, J. Math. Phys., 4 (1963), 924.
[16] E. CALABI AND L. MARKUS, Ann. Math., 75 (1962), 63.
[17] J. W. MILNOR, *Morse Theory*, Ann. of Math. Studies, Princeton University Press, Princeton, N.J., 1963.
[18] E. H. KRONHEIMER AND R. PENROSE, Proc. Camb. Phil. Soc., 63 (1967), 481.
[19] J. L. SYNGE, *Relativity: The General Theory*, North-Holland, Amsterdam, 1960.
[20] J. LERAY, *Hyperbolic Differential Equations*, Princeton (I.A.S.), 1952.
[21] A. AVEZ, Inst. Fourier, 105 (1963), 1.
[22] H. J. SEIFERT, Z. Naturforsch, 22a (1967), 1356.
[23] R. GEROCH, E. H. KRONHEIMER AND R. PENROSE, Proc. Roy. Soc. Lond., A327 (1972), 545.
[24] B. CARTER, General Relativity and Gravitation, 1 (1971), 349.
[25] S. W. HAWKING, Proc. Roy. Soc. Lond., A308 (1968), 433.
[26] R. PIMENOV, *Kinematic Spaces*, Seminars in Mathematics, V. A. Steklov, ed., Mathematical Institute Bureau, New York, London, 1970.
[27] A. D. ALEXANDROV, Voprosy Filosofii No. 1,67 (1959).
[28] R. W. BASS AND L. WITTEN, Rev. Mod. Phys., 29 (3) (1957), 452.
[29] R. GEROCH, J. Math. Phys., 8 (1967), 782.
[30] ———, Ibid., 11 (1970), 437.
[31] A. K. RAYCHAUDHURI, Phys. Rev., 98 (1955), 1123.
[32] A. KOMAR, Ibid., 104 (1956), 544.
[33] S. B. MYERS, Duke Math. J., 8 (1941), 401.
[34] R. H. BOYER, Nuovo Cimento, 33 (1964), 345.
[35] J. L. KELLEY, *General Topology*, Van Nostrand, New York, 1955.
[36] H. S. RUSE, Quart. J. Math. Oxford, 2 (1931), 190.